EARTHWORM SELLING AND SHIPPING GUIDE

by

Charlie Morgan

EARTHWORM SELLING AND SHIPPING GUIDE

by

Charlie Morgan

Revised edition

Third Edition March, 1972

Printed in the United States of America

HOW TO WRITE A CLASSIFIED AD

No matter what you have to sell, how expensive or cheap it is, others must know that you have it, and you must convince them that it is desirable. You could have a house full of silver dollars and be willing to sell them for ten cents each, but unless others knew about it, even this great bargain would be without buyers.

In the bait business there are various reasons for a person wanting your product, but the chances are that he will want to sell it at a profit, use it for breeding stock, or amuse himself by catching fish. Very few fishermen who use live bait will search the classified columns for a source; although many of them read there for amusement or to see what is new or possibly with a view to raising his own bait. The average fisherman will depend on his bait house or sporting goods dealer to lay in this commodity just as he would fish hooks.

We must appeal, then, to the operators of bait houses and sporting goods stores or a variation of one of these, because bait is likely to be sold anywhere, including wayside mechanical or "honor" system vendors. Or we must appeal to a middleman who supplies a number of bait dealers or vendors. A third prospect is the man or woman who wants to become a worm grower.

The profit motive is the most desirable appeal in bait and breeder worm advertising. This narrows the best methods of appeal to a dozen or so "leaders," or words CAPITALIZED by publications that accept classified advertising. I might say here that NO commercial worm grower has ever found it profitable to use space advertising, rent or buy mailing lists, except possibly the lists of another grower, or to use any method except classified advertising to find customers for worms.

Not all magazines have classified columns or accept advertising by the word. Those that do generally give no display space (there are some exceptions to this) and usually only the first word can be in capital letters or bold face type. We must choose this one word carefully, as well as the phrase or sentence of which it is the beginning. We have no drawings or blocked words or phrases to capture attention, so we fall back on those words known to be effective attention-getters.

Words and phrases are used slyly, boldly or teasingly, as the case may be. Many advertisers use classified ads as direct selling messages with prices and terms. Others offer literature that, in the long run, can do a better job of selling as it gives the advertiser a chance to use the graphic arts not permitted in classified selling. Pictures, cartoons, drawings, even blank spaces, are used to good advantage in sales literature.

Of the thirty or forty words most useful in selling worms directly through an ad, perhaps five are repeated most often, either as leaders or in leading phrases. They are: LARGE, JUMBO, SPECIAL, PREMIUM, and BEST. These words are followed or surrounded by not more than three or four others to make up an eye-catchind leading phrase or sentence that is calculated to make the reader want to learn more about this stupendous offer.

A "leading" word is not necessarily the first word in the opening sentence. For instance, "handpicked premium" is used for all kinds of worms and crickets. "Best handpicked premium" is used effectively. "LARGE premium worms handpicked." "SPECIAL large bait worms." Remember that only the first word, and in a very few cases the second, is or can be capitalized in a classified advertisement. Some magazines now accept complete ads to be run in **boldface.**

Several words that have proven effective when used alone or in a sentence are: DELIVERED free; quantity DISCOUNTS offered; BIG bait worms; BETTER bait and service; OUTSTANDING bargain; HUGE supply; GUARANTEED quality; QUALITY; TERRIFIC bargain; VALUABLE free instructions; FREE with order (refers to any service or additional product that might be valuable to the customer). And there are others: COLORFUL; EXCELLENT; EXCITING; IMMEDIATE (delivery); IMPROVED; INSTRUCTIVE (literature); LIBERAL (overcount); REDUCED price; and so on.

Many advertisers have several ads which are interchanged at regular intervals or monthly or seasonally. Others run the same ad year after year. My own ad, with slight variations, of 14, 15, or 16 words, has appeared practically unchanged for more than fifteen years. It does not try to sell anything, it merely invites an inquiry.

Ad writers, and especially ad salesmen, like to say "the more you tell, the more you sell." This is not necessarily so. Many items simply cannot be described and delivery terms expressed in ad space that will sell the merchandise at a profit. Worms especially are in this category. The price per single thousand may be added, but the details of quantity discounts on bulk, cupped, bait, breeder and bedrun worms together with delivery methods and terms would simply require too many words or too much space, making the ad uneconomical for the small business man.

The brevity that makes classified advertising so profitable makes a closing phrase or sentence almost as important as the "leader". We know that we cannot tell the whole story, so we put in as much as feasible and add: "Write for free literature," or "Send stamp for further details". Some advertisers have offered "Literature 10¢" or other nominal charge--but any charge whatsoever for a sales brochure or descriptive bulletin will reduce returns as much as 50%. Even requesting a stamp may reduce the returns to below the profit point.

In requesting small coins and stamps, most people say that they are "trying to keep away curiosity seekers". We spend as much as $50.00 to arouse the curiosity of a potential buyer then kill it for a stamp or a dime! You are trying to project the image of a businessman, not act and sound like a man who robs piggy banks.

Offers of free samples are not good business in the worm industry because of handling and shipping costs of live animals--and because the classified ads are read by enough fishermen planning a trip in the near future to ruin the season for you. But you can offer a "generous" sample for a nominal sum. This is a proven moneymaker, and makes a good closing phrase for an ad that might mention a much higher price per thousand in a preceding sentence. "Generous sample $2.00 postpaid," is a good example in the case of the "red" worms, "hybrids," "wigglers," "stripers," etc. Or "Generous sample postpaid $3.00," in the case of African nightcrawlers. The sample should consist of about 100 worms in each case. A small carton can be quickly cut to fit one pint cup surrounded by shredded paper for protection from bumps.

Generally earthworms, or any kind of live bait, are shipped prepaid, via parcel post, with live delivery and count guaranteed by the grower. Of course other means of shipping are used and will be covered. Guaranteed live delivery and count has been the policy in the bait business since its inception, and this should be stated one or more times in all literature and advertising where worms are offered for sale.

At least six or eight worm farms at various times have advertised as "world's largest" or "largest in the south," or "midsouth's largest." In the case of bait farms, such claims are of extremely doubtful value--especially when two or more make the same claim in the same classified column. Age in active business or continuous time in operation undoubtedly has some advertising value in establishing confidence, but nothing succeeds like "Satisfaction guaranteed" in the ad and in the literature. An advertised guarantee, or one traveling through the mails, is understood by most prospective buyers to have the backing of the Federal Trade Commission or the Postal Department, or both, and the guarantor is subject to severe penalties if he does not perform.

But Federal agencies work both ways. Simply because they enforce a guarantee or investigate questionable advertising does not mean that the buyer does not also have some responsibility in the matter. It is up to him to have proof of damage in transit or arrival in unsatisfactory condition. Or the merchandise must be returned to the sender in good condition. It is almost impossible to prove untrue advertising in the case of worms. A "jumbo" sized "red wiggler" of the Alloloborpha species may be less than three inches long, or even not more than two inches long--so they are tiny compared to other species of worms. A "large handpicked" breeder worm may be less than two inches long in one species while another species may grow to eight or ten inches or longer.

Presently the worm industry is plagued with exaggerated claims and misleading words that actually are not intended to be so in most cases. In the cool spring weather, growing conditions are at a maximum and the worms are large and beautiful. The grower whips out glowing ads and sales literature describing his lovely large worms...and the orders come rolling in-and four or six weeks of heavy picking and hot weather leaves the remains sad pigmy replicas of those so heartily offered as "giant" bait worms...and still the orders are coming, many of them repeats from happy and satisfied customers. What can or should be done? Alas! Honesty and greed, good intentions and need, clash. Sometimes one can spread the remaining worms into new beds and grow them rapidly to good size, usually not quickly enough or large enough. Sometimes a source can be found for buying good worms from small growers that are nearly always around if they can be located. Some growers make advance arrangements with one or more small growers. Sometimes the problem is impossible to solve and money must be returned.

Should one invite visitors? This was once thought to be a real drawing card. Unfortunately, most visitors are tourists and sightseers, not worm buyers or even prospective worm buyers. And visitors require someone to show them around and explain things. If this is not done they will approach busy pickers who are counting worms, and not many pickers can carry on a conversation, count and grade worms at the same time. Of course, in some cases buyers make arrangements to pick up loads of worms by truck or car in order to save transportation costs, time, or to avoid gambling with bad weather in spring or fall. But even a good customer can become quite a nuisance when checking the count and size of worms under the picker's nose.

So how far should we go with advertising that may cost up to $3.00 or more per word? $3.00 per word is higher than the average, but some of the best cost even more. Many use small teaser ads offering a "special" designed to get the inquiry. After the first inquiry and small order comes in, the name is bombarded with "leader" offers of instruction books and large breeder or bedrun worms. Non-buyers are followed up with less expensive offers of "starting stock" and possibly "free" instructions, though usually the latter is not good business as a "free" sheet or folder that can tell only a small part of the intricacies of operating a worm farm may do more harm than good. Usually the third letter is not considered a good gamble in the worm business, but a few people do mail to the same list over and over again, picking up a customer now and then. My own method is two letters with some price reduction or "special" offer in the second. The prospect will read your ad in the same or another space, renew his interest, if it can be renewed. Direct mail advertising is too expensive for earthworms.

So what we need and want is fast action. There are a certain number of "gimmicks" used in classified ads (the words quoted above) and sales letters that follow, designed to cause the inquirer to BUY NOW!!! "Time limits" are not usually effective in the worm business, while "limited supply" might actually drive customers away, giving them the impression that you might not be able to meet their season's requirement. Then we stress, if possible, other things: "year round delivery," "satisfaction and service guaranteed," "prompt delivery any time of the year," "free overcount, etc.

Anything offered free will arouse the curiosity and sometimes the resistance of customers...on the whole it is a good reaction. 75% of the people interviewed for this chapter claimed to be "terrified" of "bargains and free offers," but a search of my files shows that almost 99% of them have been full or part time customers of some worm grower for a full ten years or more. "FREE folder describes," "EXCITING details free," "EXCITING facts free," "BARGAIN list free," "FREE book" (if you have anything printed that may legally be called a book), "ITS free! Write!" Remember that legal definitions of a book vary from the Postal Department's 22 or more printed page dictum to Webster's "anything consisting of two or more pages bound together."

Generally, in selling worms by mail, we advertise for inquiries rather than direct sales, but the advertised price of a sample or single thousand might bring in enough direct business to more than pay for the ad-and we get a shot at the big dealers by adding the phrase (if we have worms) "quantity discounts" or "dealers wanted."

Many periodicals accept classified advertising at rates from 10¢ per word to $6.00 or more for space buyers who first sell classifieds then buy space in national magazines to run the smaller ads in a cluster. Generally, within reason, the higher priced ads reach enough more readers to make the additional cost worthwhile. The most serious decision for the ad buyer to make is which publications are most desirable from a readership point of view. Obviously one would not advertise earthworms in the sophisticated fashion magazines, or the younger set readership of Seventeen magazine. On the other hand, House and Garden might be a good "seasonal" medium because many gardener-fishermen buy one or more thousand worms to stock a compost pit or pile, or to keep handy in the basement for quick availability. And many gardeners buy cheap bedrun worms by the thousands in the spring for soil conditioning.

All farm magazines and publications circulated mainly in rural areas may be good advertising mediums. The widely circulated "home mechanics" are outstanding sellers of everything from worms to rare books. Popular Mechanics, Popular Science and others have a Live Bait section, and their combined readership may be as many as fifteen million people or more--not all of whom are going to read your ad, of course, or answer it if they do, but those who do spend a stamp and time will be good prospects.

Out best efforts should be aimed at the "big three" of the hunting and fishing world: Outdoor Life, Field and Stream, and Sports Afield, not necessarily in that order. These are read by the very special segment of people you want to reach and who may want to reach you...and our most complete ads should be reserved for them. And ads in any nationally circulated periodical will bring offers from other worthwhile publications-and many not so worthwhile-who will solicit you for your advertising business. This is when you must get selective indeed to make your advertising dollar count. Generally the nationally circulated publications are best, but some of those of limited (regional) circulation will pay their way for seasonal (spring and summer) ads, and possibly for year round. These are the various state farm magazines, co-operative trade papers, sectional sporting club sheets and others that you will get offers from as your activity is noted by advertising managers who scan their competition.

Should we "key" our ads? That means putting in an additional word, number, or letter, different in each publication, so that we can trace each inquiry to the advertisement that attracted it. I use my postoffice box number followed by a letter or a number, but my small town postoffice has less than a thousand boxes, so an extra number or letter is not noticeable. This may not be permitted by some postmasters, but it is not forbidden by postal regulations-yet. Some growers use different wording in addresses, such as Morgan's Worm Farm for one magazine, Morgan Farms for another, Morgan's Wormery, Morgan's Worm Gardens, etc., but this could get to be a real test of composition if one advertised widely.

I keep a small ledger, nothing fancy, in which each letter or number has a column, with daily entries for the number of inquiries from each key, and subsequently, the number of sales from each publication. These are checked quarterly, or more often in the case of a trial in a new publication, and if they are not paying I consider the possibility, first, of changing the lead word in the ad, changing the ad, changing the classification (from Live Bait to Business Opportunities, or vice versa) or dropping it altogether. If a bait or breeder worm ad does not pay during the months of April through August, it is not going to pay, period. Earthworm instruction books sell best in the period January through June.

Many publications will pay better than average in one season and be a total flop in others. What these may be can only be learned from experience, as it will depend to some extent on the type of worm you are offering (red wigglers, red golds, hybrids, brown nose, etc. can be shipped all year, while the African nightcrawler is limited in all but the deep south to summer or warm weather; the wild worms are limited to cool weather or specially equipped refrigerated transport facilities).

There is no sure and certain way of writing a profitable advertisement of any kind or length. Don't load it with useless words. Obtain copies of magazines at a newstand and study other ads in the category you are interested in. You should not copy other ads word for word but you can shuffle the effective words around in your ad and make them serve your purpose without plagiarising another writer's work, which is illegal if he chooses to prosecute. Write the publication for word count and classification information--be sure and state clearly what classification you want; I have seen bait ads in help wanted columns.

There is always a delay of one to two months, or even more in some extreme cases, from the time you order your advertisement and its appearance in the magazine. In seasonal advertising, it is vital to allow for this time lag. Be sure to get your ad in early enough for it to run during or before the first month of the season. It is well to remember that generally year round advertising pays, even on seasonal items, with possibly an expansion of the ad during the most active part of the year.

Another thing that I will emphasize: If you are going to spend even $10.00 a month on advertising, get an agency to handle it. There are advertising agencies who specialize in handling the small account, and they can do a much better job of anticipating your needs than you can...and they will be helpful in many ways besides preventing an error in timing that may cause you to miss a crucial deadline in submitting your copy. I have used Morlock Advertising Agency, 188 W. Randolf St., Chicago, Ill. 60601, for 15 years. They specialize in the small budget classified field.

THE SALES LETTER

After you get the inquiry, what?

One thing to keep firmly in mind always: Whatever you do, DO IT PROMPTLY! When an inquiry comes to you the chance is good that the same person has written to one or more other people who offer the same merchandise or service. Everything else being equal, the first letter in return will get the sale. In mail order, nothing pays off like getting there first with the best.

A neat, concise sales letter is not for the average inexperienced worm grower to write without careful thought and, if possible, some professional help. Every word of this letter is important. From the color of the ink used to the width of the margin, some impression is made on the reader--and neatness counts. It is a fact that in the worm business green ink on white paper will get 7% fewer returns than black ink on white paper--and that red and black (paragraphs or sentences to be emphasized, in red) will get 5% to 6% more returns than black alone, while red alone will get fewer than black but more than green. Stay away from exotic colors in inks and papers that may brand you as an eccentric in the minds of some readers. Plain black on white is and has always been, in style.

An order blank is useful and it may be a different color of paper than the sales letter, but not necessarily so. If more than one type of worm is being offered, it may be useful to use different colors of ink to distinguish ordering sections for each kind of worm. One school of thought has it that a light pastel color such as goldenrod induces the customer to buy: but statistical evidence points out that it will not pull enough additional orders to pay the 15% cost added for the use of colored paper by printers.

Apart from paper and ink colors, neatness counts. A well proportioned letter, not crowding the margins of the sheet, draws best. It is better to use both sides of a page, or a two page folder, than to crowd one side or one sheet with small type and close margins. If the letter is too long for one sheet, but does not fill tow, the bottom of the second page can be used as an order form by decorating it a bit and perforating for easy removal--and this may be a good place for the use of red ink.

A spread of two sheets (a two page folder) may give you a chance to offer a "special." A "fisherman's special" of a smaller number of worms in a smaller package than the usual minimum bulk pack, such as 6 cups, of 200 worms each, or 4 packages (quart cups or special bait bags--see packages and packaging instructions) of 250 worms each, are good selling items, as many fishermen will answer your ad with a view to stocking a small bed and raising their own bait, or to take on a vacation fishing trip...and many dealers and larger customers will take this means to see a sample of your product when they might not want to buy a larger quantity. Such a small sale has for me resulted in some spectacularly large dealings. As this is being written I am dealing with a man who has purchased $3.650.00 worth of worms during the past four months-his first purchase was a lonely thousand worms, with no hint that he was a dealer. By this you can see that it pays to treat every customer as a man who has the potential to make your fortune because, collectively, they do have.

"Beginner's specials" are popular items, even for established worm dealers. This may consist of one to five gallons of "bedrun" worms, together with instructions, or maybe a large single container such as a 5 gallon paint can or lard can, with 5.000 to 7.500 worms and capsules. These can be tagged and shipped through the mails.

Here again, in the sales literature, we come to "leaders" or words, even whole sentences, **to be in bold face type** (or even in red ink) to lead the reader into paragraphs and sentences, or to emphasize words or phrases. Writing a sales letter that sells is a more technical matter than writing a classified ad that pulls inquiries. And sales letters may require instant change to conform to competition. I have burned a new printing of 10.000 sales letters and price sheets, costing more than $200.00, simply because a competitor's price or letter made my offer non-competitive and outdated. These things are not easy to take, but they happen in any business, and we must fact the facts and hazards. Of course, with much care and thought, these losses can be avoided. One gets too sure and certain, sometimes, and learns that he was not as intelligent as he thought he was. Careful analysis of costs in advance of writing and mailing literature will make you certain that you are offering the best possible bargain, no matter what your competitor might do.

Starting at the top of the first page of the sales letter, we set up some kind of letterhead. This may be anything from a simple block letter name and address to a flowing color design on a pastel background. Designs for letterheads are easy to come by; your local printer is likely to be an authority, and will have cases full of type styles, and a head full of good ideas--if he is a good job printer.

Avoid the use of many-worded, tongue-twisting names for your farm or enterprise. Maudlin or sentimental names may, in this day of cynical professional and semi-professional readership, actually drive customers away rather than attract them. And many times the sentiment is understood only by yourself. A friend of mind, a physician with brains and business experience, went to the trouble of registering and publishing banns (unnecessary and not legally required unless you are incorporated-or possibly by county or local law) on a project named "End Of The Road Enterprises Limited." Imagine his surprise and disgust when he started his classified advertising campaign to find that each word in it cost him just as much as each word in the body of the advertisement itself.

A straightforward approach, in simple language, will sell more worms. Most, but not all, of the people who read your ad and sales-letter will already know something of the subject. The "new worm" pitch has been worked to death. The poultry farmer knows the species of his hens, the pig fancier knows breeds at a glance, as do growers of all kinds of live stock--except worm farmers. Strange names of all kinds are forever popping out of the skulls of worm growers--and not one in a hundred knows the species of worm he is raising. So it is probably best to put a common and well known name to your product. The person who sold you your breeding stock cannot object if you use the same name. After you study the background and history of earthworms, you may be able to come up with something distinctive, but not misleading, for your worm. You must remain within the bounds of truthfulness or be up against Federal regulations (see chapter on licensing requirements and regulations).

Ten or more different names are presently in use to describe four different species and one sub-species of worms raised and sold commercially. Only the sub-species might legitimately be called a "hybrid" and that is stretching the truth some. But all five are referred to at present as "hybrid," "red hybrid," etc. "Improved English," "Egyptian," and so on are often offered on the bait or breeder market. The "African nightcrawler" is a distinctive large tropical worm which did originate in Africa but is now found in all tropical and sub-tropical climates except the deserts--but one hardy imaginative advertiser has come up with "hybrid red African nightcrawler." It was bound to happen, and may be typical of the free enterprise system...I've said it before: The average successful worm farmer is an extremely individualistic type.

And a new one has just popped out--the "grey nightcrawler," the "wonder-worm." It is grey but is not a nightcrawler. And it is NOT a good hot weather worm-nor does it propogate readily or grow rapidly. It is common in various parts of the United States, especially the deep south, in the wild state. They are best stored and propogated in mild refrigeration, about 65-70 degrees.

Will the words "lively large bait worms (or fishworms)," result in as many orders as "lively large hybrid bait worms?" The matter is debatable. It is true that the word "hybrid" has, through much publicity in farming and the livestock industry, come to mean an animal or other farm product of outstanding size, quality and productivity. The word may be biologically meaningless, but it carries force, when use properly, and truthfully.

The words "SALE," "IMMEDIATE DELIVERY," "24 HOUR SERVICE," "SUMMER (or fall, winter, spring) BARGAIN," "SPECIAL THIS MONTH," FREE 10% OVERCOUNT" (standard in the worm business anyway--but be sure they get in there if you offer them), "FREE DELIVERY," "POSTPAID TO YOU" (both standard), are some of the key words found in classified advertising as well as sales literature.

If you have a true bargain to offer, and the product worthy, you do not have to be an advertising copy writer to put together an effective sales letter. It should be checked for reasonably correct grammar, and the spelling should be right, but here again your local printer will be glad to help you if you offer him the printing business that must be done.

Keep in mind the fact that your customer is not the least bit interested in helping you--his object is selfish, he wants to help himself or let you help him, and you already know from your ad and his response approximately what he has in mind. Simply tell him in your letter that you can fill that need and give the price and delivery conditions. If your product and price is equal to that of your competition, you will get your share of the business. And as your experience grows, you will get more than your share, and possibly at a higher price as your facility with words and knowledge of people improves.

The opening of the letter should have a salutation like "Dear Friend:," "Dear Sir:" or space can be left to write or type in the name of the inquirer, such as "Dear Mr. (or Mrs.):----Thank you for writing." Nothing succeeds like the personal touch. One or more presently very successful growers of worms types in the prospect's name in the order blank, thus moving the customer one step nearer to the order without any active participation on his part.

The closing should continue the personal approach as much as possible. "Sincerely yours," "Yours very sincerely," "Cordially yours," followed by a real or facsimile signature, depending on the size of your operation and the number of letters you would be required to sign.

The letter should definitely close with a postscript below the signature (PS.), reminding the reader again of some salient point in the letter and urging action NOW!!!

One of the classic arguments among classified and mail advertisers is about a message on the outside of the envelope bearing the sales letter. In some businesses, perhaps. In the worm business, no. First it labels your letter a mail order proposition--which of course it really is, but we'd rather it appear not to be, even in answer to an inquiry, if possible. The mails are flooded with unsolicited propositions and yours might go into the waste basket with the rest. An ordinary return address, and here a facsimile written reproduction will do a real service, is dignified and effective. And no matter how much money it is going to save you in labor or other direct costs, AVOID THE USE OF GUMMED LABELS AND POSTAGE METERS: Nothing more quickly brands a piece of mail an advertising proposition than a gummed label or postage metered stamp.

By all means use a business reply envelope for your customer to answer. This again gets your selling job one step closer to the mail box without any action on the part of the buyer, and may increase returns by as much as 10% to 20%. It can make the difference between a profit and a loss in your business. See section on licensing and permits for information on permits to use business reply envelopes.

Stuff another order blank in the package with the worms, or acknowledge the order by first class or air mail and give shipping date and other pertinent information. Any personal letter or signed letter in the package will require extra postage and a notice outside the package. But a simple printed order blank with advertising does not require extra postage when sent parcel post. I use air mail for acknowledging, with a new order blank, possibly a "special" offer, because a perishable package is moved through the mails with all possible speed, and the worms might arrive ahead of a first class letter.

Since the 1970 Postal strikes and slow downs, the federal postal service has deteriorated, especially on perishable packages. You must do a real packaging job now in order to get your worms to the customer alive, and many worm growers are shipping only to the fifth postal zone from their area of operations.

But our own farm packagers have worked out methods of packaging worms for delivery to any point within the contigious United States by adding a bit of moisture for longer desert journeys, or deleting a bit for shorter journeys that end closer to home. Nevertheless, losses are greater because handling is less careful than in "the good old days."

A few growers are using Greyhound parcel service (Greyhound Buses) or United Parcel Service (UPS) wherever possible. The opportunities for these are limited, however, because nothing covers the United States like the United States Mail service.

LICENSES, POSTAL REGULATIONS, ETC.

No matter how high the intelligence of the American citizen, he is surrounded by people who know more in certain cetegories than he does. When in doubt about anything, find an expert and ask. On postal matters, ask your postmaster. On licensing requirements, go to your county tax collector and city clerk, if you live in a town. Always, you have recourse to the County Agricultural Agent, who is certain to be able to help you.

Licensing or permits of any kind are not required on a Federal level, or on a state level, except sales tax or other local tax regulations for retail sales only may apply, for raising, selling and shipping worms. A *few* states have sales taxes, sporting goods or recreation tax, or taxes on the sale of all types of live bait; some at the wholesale level, some at the retail end. But no tax applies to interstate sales--those baits sent interstate by any transportation method. Check with your local fish and game enforcement officials or the sales tax official in your area.

Occupational licenses or permits may be required in some counties or cities for the operation of a retail bait stand. Worm raising itself is considered a farming activity by the Federal Government, at least for tax purposes (including Social Security and Old Age Benefits withholding, but not income tax withholding), and this carries over to the state and local government. And on the local level farmers are the least controlled of all businessmen. My surveys have revealed that usually state and county officials never bother about an activity in the worm business, because it is inconcievable to an inexperienced man that you may be conducting a very well-paid business.

Smal! areas at various times, are "quarantined" because of local infestations of insects that may spread and damage crops in another area. The most common of these in the south is the white fringe beetle, and in the south and southwestern citrus belts there is the never-ending fear of fruit fly and nematodes. Worms going out of one of these areas and into another, either east or west, should be labeled "earthworms packed in Canadian (or German) peat; noxious weed and nematode free." Your county agent or state agricultural representative is empowered to inspect your premises and give a clearance if there is any doubt. This service is free in most states but in some there is a modest charge and you will have to pay for the printing of clearance labels if needed. Cases of infection of any kind are rare in worms, though greenhouses and nurseries are under constant supervision. Anything from the ground moving from southeast to southwest *is especially* suspect.

Some limitations are imposed by postal authorities. In the state of Florida at this writing there is a circular in effect: File No. S6: JHC: 101, District Operations Manager, Jacksonville, Florida (since moved to Atlanta, Ga.) directing postmasters in Florida not to accept packages of live earthworms for either COD (Collect on Delivery) or insured parcel post. And the same letter further recommends that postmasters "consider" requirement of "special handling" on all packages that will be in transit for more than 48 hours. Some postmasters are evidently not aware of this letter and do accept packages for COD and insured delivery. Claims, however, are processed in Atlanta, and they are aware and do not pay.

A MEMORANDUM: Postoffice Department, date May 24, 1965, Postal Services Center, Miami, Florida, File 2MR:EP:fm 162 modifies the above definition to: The Department has confirmed the acceptability of earthworm shipments as insured matter; "Postal insurance is payable for the loss of insured packages containing worms, but it is not payable for the death of worms unless it is definitely established that death was attributable to the fault of the Postal Service." This places the burden of truth on the shipper or receiver and makes the Postal Service sole judge of its performance--which makes it very unlikely that many claims will be accepted or paid. The two paragraphs above apply to Florida only.

Parcels shipped between postoffices are restricted as to weight and size. First class postoffices accept packages weighing to 40 pounds to local, first and second postal zones: any distance beyond that has a weight limit of 20 lbs. Second, third and fourth class postoffices accept weights to 70 pounds, which is far above any reasonable number of worms that should be packed into one shipping crate or carton--though some established and successful worm farmers do it by shipping one large carton of several strong corrugated boxes packed and fastened together with tape.

There is a passage in section 135.31B; Weight and size limits of the Domestic Rates and Fees Manual that some growers make use of. It reads, in part: "Except that parcels mailed on or addressed for delivery to a rural or star route, parcels mailed at the postoffice (first class) from which served by patrons located on rural or star routes, or parcels containing baby poultry, nursery stock, etc. are subject to the limit set forth in 135.31A." Section 135.31A is that section permitting you to mail a package of farm products weighing up to 70 pounds. Your worm farm is treated as an agricultural activity by the internal revenue service, and should be covered by the "etc." to which such products are referred in the mailing section. This means that growers in or near a large urban area can get around the 20 pound limit if they live on a rural or star route. This gives you two outs.

But here again, nothing succeeds like the personal touch. It will pay high dividends to be on good terms with the postal employees at your local postoffice, even at some inconvenience to yourself.

Labeling is not exactly standardized by the Postal Department, but certain minimum requirements must be met. First and foremost, many of us are learning the hard way that if labels do not contain the ZIP code number of the addressee they are not likely to arrive in good condition. Fourth class (parcel post) packages must be labeled "Contents merchandise: May be opened for postal inspection if necessary." Or some form of this permissive statement may be substituted. And "PERISHABLE--LIVE WORMS, protect from extreme heat or cold" in large red letters on a separate label adjacent to your address label is almost imperative. Just printing it on the box is not enough, it should **STICK OUT!!!** preferably in block letters. I have shipped possibly 250 million worms in the past 20 years and have never insured any except when required to do so by a customer, and I am sure that I have not lost more than

250.000--one percent of the total. Insurance, if available and collectible, would have cost up to 5%--more at present rates.

The Postoffice Department issues a bulletin, available free at your postoffice, that gives instructions for labeling and packaging--but your valuable perishables should be packaged and labeled far better than minimum Postal Department requirements in order to protect yourself and better satisfy your customer.

A business reply permit may be obtained free from your post master. This assigns a number and allows you to include a business reply card or envelope with your mailings which your customer can use to send you an order without himself going to the trouble of finding an envelope and stamp. A printed standard form shows postal department regulations as to printing of these envelopes and their use. They may be printed by your local printer or, if your volume justifies it, you may purchase in quantity, at a vast saving of money, from wholesale paper supply houses. This is a cheap method of promoting business because you send an envelope that may cost less than a third of a cent and only have to pay postage if it is returned. I use air mail exclusively for this.

Bulk mailing permits may be obtained for a small yearly fee plus a postage fee for each piece. This permits you to mail unsealed, or "open for inspection" envelopes with sales literature inside at a much lower rate than is required for first class. Your returns will be far lower and there are certain regulations that you must follow, but it might pay. Many people use it, I never have. Your post master will explain the procedure.

Other than those listed here, there are no restrictions of any kind at present. Various communities are rigidly restricted as to commercial and farming activities within urban or sub-division limits. These may actually be so stringent as to encourage planting of climbing squash vines as ornamental but forbid picking them for food. Generally, if your premises are clean and your neighbors not annoyed, you are able to raise worms nearly anywhere you live...in fact, unless you advertise it locally, your immediate neighbors need never know about it.

HEY!

MR. POSTMAN: KEEP ME COOL —IF I GET TOO HOT I'LL DIE, AND IF I DIE I'LL STINK LIKE HELL, BESIDES SOME FISHERMAN WILL BE MIGHTY UNHAPPY CAUSE I CATCH BIG FISH.

MR. POSTMASTER: I'm just merchandise, but if you want to see some mighty fine bait worms just open me up for postal inspection.

FROM: Charlie Morgan Box 116, Bushnell, Florida

To: ....

....

....

Going Fishing! FROM:

RESEARCH ASSOCIATES,
P.O. Box 712,
Bushnell, Florida, 33513

CONTENTS: MERCHANDISE.
return postage guarantee
May be opened for postal inspection.

TO:

No. worms ______________

No. Pkgs ______________

Ship. date ______________

©1967
research associates

The labels above are examples of some in use for many years. There are literally dozens of different designs, some primarily for advertising the worm business and to warn the mail handlers, while others are primarily to warn mail handlers. It is illegal to copy another grower's label, copyrighted or not, because he has established prior right by prior use.

Recently, new international postal regulations have changed between the United States and Canada. Formerly, we could ship worms by parcel post from the U.S. into Canada and they would accept them there for delivery, even though their own regulations forbade worms to be mailed in Canada. Now, they cannot be shipped into Canada by parcel post from the United States. They will be turned back at the border and you will likely lose the shipment altogether for they might not last long enough for the return journey--collect postage, of course.

The only reliable method of delivery into Canada is by air freight. Air parcel post has the same regulation of non-entry, and we have found that REA Railway Express Agency is simply too slow in getting there with the goods. Of course if you live near the border, or your customer lives near the border, the problem is reasonably easily solved: You can ship to a friend on the U.S. side and he can pick them up there and make his way through the port of entry by filling out a form or two and paying a small customs fee. Or he can pick them up at your place.

When shipping by air, try to remember to ship collect. Give the customer something off in the manner of a rebate or discount for his expense in paying freight. *The reason for shipping freight collect is as simple as spending a dollar–the airlines know that a collect shipment of dead worms is not only going to be refused, but it will not be paid for and must be returned, at their expense, to the shipper, if the consignee (receiver) requests it. SO THEY WILL HANDLE YOUR WORMS WITH EXTREME CARE under these circumstances. Airlines are usually good at settling claims, too, if you can prove that you have successfully shipped worms for some time without more than the usual problems.*

SELLING WORMS

A home retail worm business is as simple as putting some worms into a container, putting up a sign in the front yard, and perhaps running a small classified ad in the local paper. The packaging need not be fancy or even careful because here we are not involved in long distances nor do we trust our worms to the hands of unsympathetic mail handlers or unknowing dealers.

Fifty or 100 worms is the rule, with various other quantities such as one, two or more dozens being the exception. Nearly all bait dealers and growers add 5% or 10% overcount because, even in the most careful operation, a few worms will be injured and die or be undersize-or just plain "disappear," and the package may be anything from a reclaimed and cleaned (thoroughly, in hot water) motor oil can to a fancy printed snow white bag or plastic or wax coated kraft paper cup.

The small retail packages are put up fresh (see chapter on preparation and use of packaging materials) daily or even more often as needed and weight is no problem because they are not being shipped or even hauled any great distance except possibly by the customer. The only concern is that the material be damp enough to preserve the worms and that the package be closed to keep them from crawling out in case they are to be kept overnight by the customer or yourself.

Depending on the type, packages may be closed by using a piece of cloth and a rubber band in the case of open top tin cans, to folding and stapling--or simply securing with a paper clip--the paper bag. Patented R.C. Bait Bags come with or without attached tin ties for self-closing. I will say here, and might repeat later on, that the ordinary brown kraft grocery or hardware bag is NOT suitable for worms because the seam and bottom fold are not secured with waterproof adhesive and they come apart as soon as the damp packaging material comes into contact. There is a bag made for ice and freezer use that some growers use, but my own trials with them have been unhappy for me and for my customers.

A local bait route in which the grower serves as wholesaler to a group of bait stores, resorts, and fishing camps requires more elaborate preparation and sales technique than home sales. The farther you get from home, either in person or by proxy (mail sales) the less profit you will earn in proportion to the dollar of sales...but the farther you go the larger your potential market becomes, thus permitting unlimited gross income and profit. At home the middleman is eliminated and you earn his profit as well as the grower's. But on a route you serve the middleman, or retailer, who in turn serves the consumer, or fisherman. Thus you must lower your price to permit your agent (the middleman-retailer) a fair profit for his involvement. It must be kept in mind that 15 or 20 retailers can each probably contact 20 or more times as many customers as you can by your own local efforts because they are likely to be located in favored areas calculated to be in the way of customers.

Sporting goods stores and the outlets for things used in fishing activities are the obvious places for fishermen to buy bait; but almost any retail outlet, including drug, grocery, and discount stores, filling stations and even some automobile showrooms, sell fishing tackle and bait.

Chances are good that one or more places in your town are presently selling bait, and the chances are even better that others would handle worms if they had a reliable source of supply without too much trouble to themselves. Probably some of the places who now handle worms would change if it were to their advantage. You, as a producer, can make it worth their while at some profit to yourself. You will hear cries of pain from the competition. But in the worm business as in any other, there is room for more. And it is nearly always the successful competitor who is the "dirty" one. You can be honest and fair and still be eminently successful in the worm business. In most cases the cries of pain will be from those who have charged all, and sometimes more, than the traffic will bear, rather than basing their charges on a fair mark-up and profit-sharing basis.

The earthworm industry has always needed competition-conscious businessmen. In our expanding economy it needs them worse than ever. Check the resorts and fishing spots within a day's drive of your town with the view of setting up a route for yourself or an employee. Package some of your worms and show them. Survey average retail package size in greatest demand and cater especially to this type of business at first, diversifying later, when you better understand the needs of your customers. If you are growing worms according to this writer's instructions (see list of best-sellers on the back pages of this book) the chances are that you are producing a better than average product, and your chances of success are therefore enhanced by your more than average knowledge of the business.

On your first call, offer the retailer a consignment stock. Have some neat red and blue signs printed "LARGE RED WORMS" or "LARGE FISHWORMS" and ask the retialers to post one or more of them in conspicuous places. Leave 12 or more packages of bait, depending on his or your forecast of potential sales, have him sign a memorandum invoice (such as small groceries and hardware stores use as receipts) in duplicate, showing date and number of packages as well as the agreed price per package, and tell him you will be back in seven days or less.

Revisit (it is important that you be *ON TIME!*) your dealers in not less than a week with an ample supply of fresh worms in fresh packages--and your copy of the memorandum previously signed. Pick up the old worms and replace them with lively fresh stock. Charge the retailer for the packages sold. Make arrangement to collect cash each time, or arrange a more businesslike transaction of monthly billings and collections, based on your knowledge of his willingness and ability to pay.

Take the old worms and return them to a bed and in a few days they will be ready for another trip to your customer's counter. In this way losses are reduced to a minimum and more sales are made to better satisfied customers than by leaving worms so long in the package that both worms and package show signs of deterioration. Everything should be done to obtain and keep repeat customers. The advertising dollar is not well spent if the customer is not kept happy and brought back for another purchase.

"Honor" vending systems have been made to work in some areas. This consists in putting a few containers of worms in a convenient location (convenient to the customer's regular line of travel) and putting up a sign asking the buyer to leave 50¢ or $1.00 in payment for each package taken. Usually the money is inserted into a slot in a door or wall to prevent pilferage by passerby or vandals. In my area there is a woman who works away from home. Daily she puts up 10 or 12 packages of worms and places them on a box on her front porch. A sign above the door requests the buyer to place $1.00 there. The can contains 100 good large worms, sometimes one kind, sometimes another, depending on the season and type available. She has earned thousands of dollars over the years and says her losses are negligible.

At different times those of us who advertise obtain circulars through the mails describing vendors of all kinds; some for bag type containers and others for cups of various sizes and kinds--most of these are never heard of again. But there are vendors available that will supply the proper environment to store any kind of worm in any climate. Prices of these vendors vary up to several hundred dollars and naturally the most elaborate is the most expensive; but even in vending such profitless items as cigarettes, the machines pay for themselves by prevention of loss by handling, shoplifting and exposure, as well as reducing or fixing storage space.

Routes requiring several days or a week to cover, with overnight stops, are not unusual and pay very well. While one or more drivers are covering such routes, pickers are at work on the farm putting up packages for the next round.

Combinations of routes and mail sales are the usual method in the worm business, and there are at least two growers of my acquaintance with routes covering more than a thousand miles. They serve wholesalers who have their own routes, as well as retailers along the way, with different markups for each. They earn an excellent profit but of course they work many more than the average number of hours during the season.

PACKAGES AND DELIVERY METHODS

The ability to package worms for delivery to the buyer is divided into several exercises in ingenuity. The simplest of all is to package for sale at home or in your own bait shop where neither wieght nor long survival in the package is a factor. In these cases we must only be concerned that they are presentable. But of course the neater and better the package and packaging materials, the more likely we are to acquire and keep customers--and repeat customers are most desirable in any business.

Always keep in mind that in doing business with people, neatness counts. It may not make a conscious impression, but you can be sure that the sub-conscious awareness, even of people who are not themselves particularly neat, notes disarrayed premises, working areas, smudgy packages and scattered, untidy materials and tools.

Attractive packaging has become so important that many of the items you buy cost much less than the package in which they are contained. Cigarettes, pressure type shaving foams, fish hooks and sinkers in plastic snap-lid boxes--the list is almost endless. Container manufacturers hire artists, design specialists, and engineers whose only function is to research and plan attractive packaging, even subjugating quality to appearance. Marketing researchers spend years and millions of dollars in studying buying patterns that may in the end result in only slight change in a package design or color, but the manufacturer gains sales and is well satisfied that he got his money's worth. I go into this only to point out that it is to your best interest to give serious consideration to using clean, well designed packages for your worms.

Unfortunately, our choice of packages designed exclusively for worms is severely limited. We have two or three types of cup and a single specially designed and patented bag. All have their advantages and shortcomings, but each is good in its particular field or for a particular type of worm.

The two or three types of cup boil down to one really effective all-purpose bait cup, and one type, with several different styles and sizes, of bag. The ultimate package is yet to be invented for live earthworms. The Sealright fishbait cup, printed with a jumping fish design, comes in three sizes; half pint, pint, and quart. The lid has lately been "improved" to prevent flying off when thrown or shocked in transit, and to prevent crimping when it is put on. Actually, this "improvement" has only loosened the lid so that if it is accidentally deformed before use, it will return to its deformed shape after installation on the package and will therefore sometimes pull itself off the package. There is one stock number for all three sizes: 841, and a separate number, 839, for the cover. This cup has been changed some from past years and may be changed more--there is room for improvement of the one offered at this writing--the old cup was better. The best source of these for the small user is Wa-Su Package Service, P.O. Box 8896, Prairie Village, Kansas 66208, or write Sealright Co., Fulton, N.Y., Kansas City, Kansas, Los Angeles, California, or Peterborough, Ontario, Canada for name of the dealer nearest you. These plus gallon sizes by the same manufacturer for bulk shipment may be purchased from local dealers in many areas.

R.C. Bait Bags are especially designed for worms and come in a great range of sizes from about one pint to five quarts, single or double wall for shelf storage or shipping, printed (with up to four lines--your name, address, etc.) or plain, and are very adaptable to a great range of purposes for which they were originally intended. It can be safely said that without some type of bag or flexible package it would be impossible to ship the African nightcrawler with present rough handling by postal employees. Write to R.C. Bait Bags, 224 Eleanor Street, Kalamazoo, Michigan, 49006.

R.C. Bait Bags are suitable for either retail over-the-counter sales or for bulk shipments. The most common bulk size is one gallon in the case of Sealright semi-rigid coated containers with slip-over covers, or five quart double wall rolled bottom porous ("They Breathe") paper bags in the case of R.C. Bait Bags.

Various other types and styles of packages show up then disappear or fall back to a limited market. Lily-Tulip, New York, N.Y., brought out a paper cup essentially the same as that by Sealright but the insert type lap-over cover was not suitable. A VERY good Lily-Tulip "misprint" is presently available from R.C. Bait Bags. How long this bargain will be available we can only surmise, and hope.

Stroud Industries of Cedar Hill, Texas offers a wax-lined cardboard container suitable, but not entirely satisfactory, for bulk shipment of worms without using an inner package. Let me say here that it is impossible to ship the African nightcrawler, or African red worm (both the same) in anything that I have yet seen without an inner package, as they will crawl when daylight shows through the tiniest opening, while the "red" worms will remain in the damp peat under most conditions.

Mutual Plastic Mold Corp., La Mirada, California offers moulded plastic containers of unspecified sizes. Another attractive plastic container is offered by Airlite Plastics Co., Omaha, Nebraska. Crown Plastics, of Fort Worth, Texas offers "Thermo-Kup" in an odd size and shape, and the last time I tested any of these the covers were offered at a much higher price than the cup. Of all these, only R.C. Bait Bags have the distinction of having been designed especially for packaging worms, and have been issued a patent (No. 2.735.215) covering the use of bags made of wet strength paper held together with moisture proof adhesives for the packaging of live worms and other types of bait. It would be interesting to know if a slip-up in quality control (use of non-waterproof adhesives, for instance) would void the patent.

Sealright's kraft cup, Lily-Tulip's waxed paper cup, Thermo-Kup and the others, even bags, started out in life as food containers, but were adapted to the use of the worm dealer. The alrger sizes of rigid containers are still used mostly for ice cream and such, only incidentally for worms.

In any case, we must attempt to package the worms in a manner that will overcome careless or indifferent handling by postal employees, which remains the best, but not the only, method of delivery to distant customers. Without parcel post there would be no quick and easy method of reaching thousands of small towns. And small packages of 1.000 worms cannot be economically sent by any other means.

Packages or separate cartons of any number weighing 40 pounds or more in total weight and consigned to one person at one address may be economically shipped by Greyhound Package Delivery (bus) or Railway Express, provided one of these has a depot or agent in the town to which you are shipping. It is well to be careful that packages are not transferred from Greyhound to other bus lines that are not so well equipped or careful in handling perishable freight. And always be certain that the buyer understands how the worms will be shipped because agents may not realize the perishable nature of live bait and make no special effort to locate the consignee or care the the animals. Be sure, too, that the telephone number of customers is placed on the label or carton, if possible, so that locating them will be easier for the agents.

I might say here that generally in the east we have had much better understanding by Greyhound claim agents than by those in the west. In most cases the eastern divisions have paid claims promptly, but not without thorough investigation, which is as it should be, of course, but in the west we have been summarily dismissed without proper hearing. These are things to be takén into consideration when shipping.

I hope that I have made the point that, in the long run, it is cheaper to buy and use a good clean, effective package, than to go out raiding garbage dumps for tin cans and other cast-offs that do a poor job in the end, anyway. The cost of packaging bulk worms for delivery in the best possible condition, considering the fact that the worms are enclosed in an extremely small space, is slightly over 2% of the wholesale price. Yet some growers try ice bags, grocery bags (entirely unsuitable) and even sew old sheets and rags into cloth bags, none of which will do the job half as well, and may cause greater expense in the long run, counting losses and unhappy customers-possibly lost customers-than the most expensive package on the market.

When it is available, worms use more oxygen per pound than does man. With a good supply of air they are lively and colorful. As the oxygen in any air-tight container is depleted they become less active and their skin turns dark as blood surfaces to pick up oxygen that is not there (worms breathe through the skin). So it is important that a packaged worm be provided with as free a flow of air as possible--provided, of course, that it is not hot, dry air that will dehydrate or cold air that will freeze. But air, even at unsatisfactory temperatures, is desirable and necessary to keep worms in top form, or return them to it after shipment.

The regular Sealright fish bait cartons have CUT holes in the lid that permit some movement of air into and out of the container. The larger ice cream cartons and other types of coated paper cartons must be punched with an ice pick or other sharp tool. This merely pushes aside the fibers, does not cut and remove them. In a short time the fibers draw together and reseal the hole. The worms usually remain alive, it is true, but they would be in far more satisfactory condition if better packaged. It is safe to say that NO worm grower has spent the time and money necessary to research the problem of packaging and delivering worms in TOP condition. It can be done. If worms are packaged properly, they will actually improve in the package.

Large rigid gallon (ice cream containers, for instance) cartons are absolutely unsatisfactory for delivering the African nightcrawler. The African will crawl out of any opening made in the container; the red worms and other types usually, though not always, remain in the bundle of wet peat. In handling while in transit, rigid containers are crimped, crushed, or deformed enough to permit worms to escape. So we should use a pliable inner package that will permit a flow of fresh air, covered with a strong shipping carton that is suitably ventilate.

The waterproof, air tight, clear and opaque plastic containers (freezer bags and storage bags of various kinds) have proven unsatisfactory because when sealed they exclude all air, and when punched for ventilation they tear or rip, and the African will force the anterior part of its body out and make the hole larger--then other worms will follow the source of air. Various types of insulated ice cream and retail packages have been tried, but lack of quality control leaves small holes at the lower corners through which worms can and do escape.

A few growers are using ice bags by doubling them; that is, by slipping one bag inside the other. This actually makes a four wall bag, as it is double walled in the first place. In testing this method myself, I found that I had more losses from worms escaping through holes in the bottom fold and dieing between the layers of paper, had trouble finding shipping cartons adaptable to the odd sizes, and used much more labor in putting the bags together (getting two obstinate paper bags to fit in and over each other snugly is not easy) than I saved in purchasing this cheaper bag.

The ultimate package has not yet arrived, but with some reservations we can recommend the S series of R.C. Bait Bags (S-30, S-50 and S-754, all double wall) as being unsurpassed for shipping worms in bulk lots. And the smaller single wall models for retail packaging provided they are used and stored properly (not touching each other) nor expected to last for more than about six days after worms are packaged in them.

Any worm that is refrigerated or package fed may be kept up to 30 days or more in the package. We are learning more and more about package feeding and will cover the subject under the proper chapter heading in this book. The coated cups are the only packages tested that lasted to the storage time limit permitted by the vitality of the worm--and of course the worms used varied a great deal depending on species, air temperature, etc.

In paper bags (R.C. Bait Bags, specifically) the worms can be removed at the end of the established period, repacked in new bags and new packaging material for another stay on the shelf. This can be done several times and the worms will still be in better condition than if packaged without ventilation. The secret is in the fact that a free flow of air is permitted through the sidewalls of the bag. The cup, having holes only in the top, must be stored in single layers for best results. The average dealer will stack them several layers high, thus effectively closing the holes and any possible change of air, by the upper cup covering the lower lid.

The plastic or wax coated carton causes the worm to decline for lack of fresh air. The paper bag deteriorates and permits the worms to escape; yet both these packages are extremely valuable to the bait dealer when used properly in conjunction with a small area converted to the use of a few stacked bins or wash tubs. Many dealers are willing to buy in bulk lots, store and package the bait worms they use in their business. A cheaper grade of worms may be purchased and conditioned and quantity prices may be used to earn a profit that will much more than pay for the time and space.

Growers should cooperate with their customers to the extent of furnishing a stated number of small packages with each unit (say 1.000) purchase of worms. The small sizes of bag or cup could be included with each bulk shipment at very little additional cost. All types of worms store well, in cool temperatues, and some will reproduce, in German or Canadian peat moss (see chapter on package feeding and conditioning). The consumer thus receives the best possible product in its freshest most desirable condition, at a minimum price to him, but at a good profit to all handlers. And the cost is lower than shipping prepackaged worms--by a good bit.

Do not attempt to raise worms in pure German or Canadian peat. The quality is not the same as it was some years ago. To my knowledge no successful worm farm uses pure imported peat. About 20% is the present limit, with Soilless Worm Bedding made from our formula, or other materials, described in books dealing with raising worms.

Routes that are serviced from a central location can be extended by using the durable, long-lasting coated cups or can be furnished a fresher product by more frequent visits with a free-breathing flexible paper bag. This at least gives the route man or grower a choice which is more than could be said a few years ago.

Other packages are on the way. The industry has grown large enough to attract researchers and investors. Bait users are able and willing to pay for a better packaged, more attractive product. Dealers are becoming more conscious of the profit involved in live bait sales, and especially more knowledgeable regarding the demands of the live merchandise and his buyers. Beer cans and rumpled paper are no longer good enough. Nylon and other plastic cloth does not rot or mildew and is porous, strong enough to withstand any handling or time element, and may be the next logical step in live bait packaging. Tests prove that worms can be stored 120 days and more without harm to either worm or package provided environmental demands of the animal are met.

SHIPPING CARTONS may be so important as to deserve a separate chapter, but we will include it here. There are few industry-wide standards. In Florida and some areas of the south, the wire bound wooden crate, which comes in a hundred different sizes and shapes, is widely used and is more or less standard in local areas.

In some parts of the midwest and west, certain types and sizes of cardboard cartons have been widely used. A standard, but arbitrarily sized, carton is presently available from Wa-Su Package Service, P.O. Box 8096, Prairie Village, Kansas 66208. My own liking is for a carton that holds tens, and multiples of tens of prepackaged bags or cups, and thousands, or multiples of thousands for bulk shipment. And this would meet the needs of all growers and shippers. Some type of carton is obtainable in any part of the country, while wire-bound crates of a suitable size may not be, or may be more expensive to have shipped in. And any kind of cardboard is lighter by far than wire-bound wood--though it is not so strong.

Many growers and shippers are using a small (Wa-Su type) carton enclosed in a larger carton cut to fit five, ten, etc. The large outside carton has holes cut in it somewhat like the "chick pullman" used by hatcheries to ship day-olds in. Nearly all the shipping cartons presently used, except by the very largest growers, and some of the smaller ones that have banded together for quantity purchases, are adapted from some other industry and are either slightly too large or slightly too small.

Some shopping and visiting of crate or box factories in your area will pay. Reclaiming and cutting to size those cartons discarded by butcher shops and other stores is fine for the small operator, but gets to be a time-consuming and expensive business as the farm's output increases. And the garbage of one business never looks good when used in another. Always try to project the image of a business man, because that is what you are, or are in the process of trying to become.

Many different kinds of meat products (boiling hens, pork loins, fish) and vegetables (lettuce, carrots, celery) are packed in wire-bound wood crates. Examine one of these and you will see that it can be taken apart simply by bending four or six wires, depending on the size of the crate, then it can be flattened for storage. Their flexibility and strength helps resist damage in handling by Postal Department employees...but it also prevents the use of rigid inner containers. The inner package should be flexible in order to retain the advantages of the strength of wood and metal in the crate. And one of the greatest advantages of the wooden crate is that the paneled open construction permits a free flow of air from any direction. It is almost impossible to stack them so that iar will not move to and through them. But a wooden crate that will hold five thousand worms is almost twice as heavy as a cardboard carton of the same size.

In small shipments we are forced to use small cardboard cartons. There are small wooden crates used for packaging bolts and heavy metal electrical components but they are as expensive as their larger counterparts, and almost as heavy. Remember that weight, no matter whether it is worms, water, wood, or peat moss, is expensive to ship.

Small shipments usually consist of 1.000 worms and multiples of that number, although some growers have done very well, profit wise, with a "generous sample" offer that consists of 100 or so worms, postpaid. Price lists should show multiples of 1.000, 5.000, 10.000, and so on in order that packages may be standardized from 1.000 worms in the smallest to 5.000 in the largest.

"Standardization" may be considered complete if you have a package to hold snugly, but not tightly, 1.000, 2.000, 2.500 (for very large African nightcrawlers or northern nightcrawlers) and 5.000 in either bags or rigid containers. Five number S-754 R.C. Bait Bags will fit into a shipping crate that measures 11x17x7 inches. 500 large nightcrawlers or 1.000 medium nightcrawlers or breeder redworms, brown nose worms, hybrids, etc. may be packed into each bag. The bags are folded and stapled at the top then placed upright in each end side by side then one on its side in the center. The crate has wire loops for fastening down the cover. And there is no better type of package for strength and ventilation. I should repeat that wooden crates are for bags and bulk type flexible packages only. The rigid cartons will not work in them.

One gallon ice cream buckets (cardboard tube cans) are the traditional method of packaging worms for shipment. Doing things in the traditional way leads to doing them for sentiment, and there is no more room for sentiment in the worm business than in any other if you have the time and money for sentiment, fine. Five one gallon ice cream cartons will fit into a cardboard carton measuring 16½x 16½x7¾ inches. This same carton will accomodate 32 Sealright 841 squat pint (16 ounce) cups. This carton may be made more versatile yet by splitting it horizontally in the center then telescoping it together to fit 16 of the same size cups. When shipping cups or small retail packages of any kind, internal dividers of cardboard should be used to separate, horizontally, the layers of cups. It is imperative that bags be separated both vertically and horizontally, as they have no structural strength and they will press together and kill the worms that are caught at the bottom.

The bottoms of cups are smaller than the tops, and pressure from a cup placed directly on the top of another cup will press out the insert of the lid of the lower cup. I lost many dollars learning this elemental lesson in engineering, thus the reason for horizontal dividers with even this strong package.

For some purposes only a cardboard shipping carton is satisfactory. Worms prepackaged in a cup can hardly be delivered satisfactorily in wooden crates because the very flexibility that makes them so valuable for use with bags makes them worthless when used with a rigid carton inside. And cardboard cartons may be obtained that will fit the exact specifications presented by your shipping problem. In fact they can now be purchased from mail dealers. Cardboard dividers and internal supports can be cut at the factory, usually from scraps and trimmings, more cheaply than they can be found and cut as needed. But they can be marked and cut quickly and easily with a "razor knife" used in grocery stores for opening cartons.

One of the problems encountered with cardboard shipping cartons is that moisture, either from packaging materials spilled from broken rigid inner containers or "wicked" from the damp packaging material in bags, will soften the kraft paper or cause it to come apart where the plies are glued together. This also causes frequent loss of labels unless a waterproof material such as sodium silicate (obtainable at any drug store in pints) is used to fasten them to the package. Sodium silicate is also known as "water glass" and is waterproof after it dries under the label. So I use crates for shipping bulk worms in bags, and cartons for shipping worms prepackaged in retail size cups.

I must repeat again that for packaging worms of all types, bulk or prepacked in small containers for retail sale, it is best to standardize. The average order of worms will be that for which prices are quoted (sample: 1.000, 5.000, 10.000, etc.), so you can help your own cause by quoting prices accordingly. The same rule will apply for prepacked retail size cups or bags. I like and have used dozens, or units of 12, 24, 36, etc. with 60 being the largest single package of cups.

The number of worms per shipping carton must necessarily be controlled by the number in each small unit. Some of the very large African nightcrawlers can only be packaged 36, or even 24, to the pint...and this applies even more so to the northern native nightcrawlers, or dew worms of Canadian origin. Average bait sized African nightcrawlers are packaged 50 to the pint (55 actual count to provide for the standard overcount). Up to 250 redworms, red golds, or stripers (we are not here recognizing any of the latter as a different species but as different names for the same-or a mixture of-worms). Half pint sizes of cup are common in some areas, and these hold 50 red worms, 18 to 24 African nightcrawlers and 12 northern nightcrawlers or dew worms for a short time. Quart size cups are available from Sealright dealers or Wa-Su.

Of the bags available, only one manufacturer has come up with a special design specifically for live worms packaged for any one of the several different purposes for which they might be packaged. The sizes range from one pint single thickness paper to 5 quart double wall heavy melamine treated paper: No. 20, pint long; No. 25, pint squat; No. 30, quart long; No. 35, quart squat. No. S-30, quart double wall shipper; No. S-50, three quart double wall shipper; No. S-754, five quart double wall shipper. These allow for a wide versatility in quality packaging for over-the-counter sales as well as for bulk packaging for shipment--but their lack of durability may result in losses when used for prepackaging for shipment where they will be several days in the carton before arriving on the dealer's shelves.

Please keep in mind that article numbers, sizes, shapes, and quality are subject to change, and all these products are in a constant state of improvement and change. The smaller sizes of bags come with or without a tin tie self-sealer, the larger sizes and plain small sizes must be folded and stapled at the top--or simply fastened with a paper clip in home or shop operation where no rough handling is involved. All can be ordered specially printed with your name, address, phone number, etc. at extra cost.

Bags are the most completely adaptable for placing in rigid or semi-rigid shipping containers. Their best feature lies in the fact that the porous paper permits free access of any air available to the shipping container from any direction. And the bags seem almost to have been designed for use with one of the world's most common waste products in mind--the empty beer carton. The carton is designed to hold four six packs of beer cans and is adaptable to half a dozen or more variations in size and number combinations of R.C. Bait Bags. Wax paper, or strips of foil, may be used for dividers and at the bottom to prevent "wicking" of moisture into the cardboard. The carton designed for 24 bottles of beer is also a "throw away" item in almost every city and town in the country--these adapt perfectly to use with the larger bags for bulk shipment.

It is important to remember that any carton, crate or box that is used as worm container be well ventilated. If custom cartons are purchased, the manufacturer should be instructed to cut four slits, two in each side of the box--more in large cartons. To preserve structural strength, the slits should be vertical, one quarter inch wide and about 3 inches long. This should not add to the cost of the cartons. The slits can be cut in salvaged cartons with a sharp knife.

There are staplers suitable for assembling and fastening flaps of cartons. These have a sharp pointed base which can be inserted through the cardboard in order to fasten top flaps. Quicker, much stronger, but more expensive in the long run, is No. 898 Pressure Sensitive fiber glass filament tape from Minnesota Mining and Milling Co., through your office supply and equipment stores. These stores also sell staplers and supplies. Either of these is much advanced over the paper tape that uses a water soluble glue which may be dissolved by internal moisture wicked out of packaging materials or by exposure to the elements in shipping. Binder twine is still used by a few growers, but mail handlers use the slack to lift the package, frequently cutting and damaging both inner and outer packages.

The usual material for packaging earthworms and some other types of live bait is imported peat moss. Peat comes in several grades and styles and is usually prepared for use in gardening and the greenhouse industry. It is only a lucky coincidence that with a little preparation it is also suitable for packaging worms. Peat moss is fine ground, for use in conditioning the soil in small planters, etc., to coarse, and unprocessed. The type used by the earthworm industry is the coarse ground or chopped kind that is pressed in bales, usually labeled "sphagnum." The full bale of Canadian peat is six cubic feet; German, 7 cubic feet. The average bale of Canadian peat moss is slightly larger in dimension than the German because it is under less compression.

Canadian peat is packaged in polyethyline-lined four ply paper bags. German peat is packed in burlap bound bales with lathe and wire binding. There are several grades and brands of each kind, but on the average one is about as useful as the other, but avoid the finely ground (almost powdered) grades which usually come in smaller packages of clear plastic. Each grade and brand varies a good bit in the matter of root and twig content and other foreign matter that may be a nuisance, but makes very little difference in the final use. On the average German peat is cleaner and more evenly ground. But German peat may vary more in acid content than Canadian. The acid content may or may not be important, depending on your method of packaging, the type of worms, and whether or not you intend to include feed with the packaging material. In the dry pack (comparatively) outlined here, for the first time in any book, acid is not important in the quantities usually found. But if feed is used, it may have a bearing on the package life, cutting even the wax coated cartons to a few days, at most.

Many things, including various tree barks, vermiculite, and other insulating materials, rock moss, sawdust, and even composted vegetable materials have been tested at various times by different growers with poor, to average, to disastrous results. On the average, the peat mosses are most easily available and deliver the most consistent results. The average cost to package 1.000 worms is about 3¢ so our instructions will deal solely with this product.

There are various methods of dealing with peat moss once the type or brand is decided upon (and don't be afraid to try a new brand if you think it might benefit your operation). Some people rub it through a one quarter inch screen in order to separate and fluff it. I have seen on one large farm a circular "squirrel cage" affair of ¼ inch screen with cross braces of 1x2 inch lumber and angle iron. This drum was turned by a ¼ horse power motor with a very small pulley with a V belt to a large pulley on the "fluffer." This reduced the speed to perhaps 90 or so revolutions per minute. A bale of German peat is placed in the drum (Canadian peat is not compressed to the point that it would work well in this type of machine), and the bale is tossed and tumbled until it feeds itself through the screen to a bin below. Small amounts of water are sprayed on the bale at intervals to soften it, more is added later to presoak it properly.

Water *MUST BE ADDED AT LEAST 24 HOURS BEFORE* the peat is used. Some writers have claimed that the use of hot water or a larger proportion of water at one time will eliminate the necesity for presoaking. In the first place, hot water is almost immediately reduced to the temperature of the peat and air, therefore losing any penetrating qualities that it may be imagined to have had; in the second place, there is no possible method of mixing any proportion of peat and water in a manner that will immediately impregnate the inner air spaces of the peat fibers with water. And it is the quantity of water in the inner structure of the peat fibers that determines the condition of the worms for the several days that they remain in it.

Peat is composed of woody fibers containing millions of cells that all plant fibers contain, more or less porous. The dried woody fibers of peat have interstices between the cells as well as a dehydrated cell structure in which to store water. These cell structures *MUST*, if peat is properly prepared, be filled with water. The filling takes time unless done under pressure. Normal impregnation time for a complete filling of the fibers is 36 to 48 hours. But a good job can be done in 24. You must plan the operation to permit the necessary time.

Peat should be presoaked in a container of some kind. Most farmers have a large tub, tank, half barrel or concrete vat in which to prepare peat. It can be broken up, placed in the container with the proper amount of water, or near the proper amount, based on the fact that about 50% of the weight of water will prepare 100% of the weight of peat-or one half to one, in weight ratio of water to peat.

Canadian peat can be presoaked in the bale by punching a hole in one end with a stick or broom handle, then pouring in the proper amount of water--slowly, so that it will distribute. But this is not entirely satisfactory because the water tends to run to the bottom of the bale. The best method is to break it up and "fluff" it with the hands, a rake or fork, then pour the water over it in a large container of some kind, working the water in and mixing the peat to distribute the water over it well. Small growers use a wash tub or other small container, but a large enough amount to last several days can be prepared at one time because it does not sour or ferment.

The presoaking does not prepare the peat completely. But it will load the internal fibers with moisture, which in turn evaporates slowly, allowing the worms to remain in good condition over a relatively long period of time. Surface moisture evaporates quickly, and may have to be replaced on the presoaked peat before the worms are packed. In particularly dry areas, or when a warm dry wind is moving the air, the peat may feel or look dry from loss of surface moisture even after it has been presoaked with the proper amount of water. This can be corrected by the addition of a small amount of water then working or fluffing the peat again.

Generally peat can be weighed more conveniently than it can be measured, especially as its volume varies with more or less compression. And we can measure water more easily. One Standard gallon of water weighs almost exactly eight and one-third pounds (8.3389 lbs.). Pints, 16 ounces (1 pound); quarts, 32 ounces (2 pounds) and so on at least closely enough for a rule of thumb mixture. And a rule of thumb that gives the best average results is to add 40% of the weight of peat moss in water for presoaking, then add another 10% or so of water when the final preparation is made for shipping.

The chart below will give the proper proportions of water to peat for shipment from and to given areas. It is impossible to predict movements by transport and temperatures in all areas. Worms shipped from some areas, especially in the south, are subject to a

Pans of worms and dirt exposed to broad daylight, where the worms quickly find their way to the bottom of the container. After the dirt is carefully raked away, pure, clean worms can be lifted out for weighing or measured for quick, cheap handling.

Bagged worms cooling out to 75 degrees F. they will be crated 5 to 10 bags (1000 worms per bag) per crate, for shipment to almost anywhere you can name in the United States.

The kitchen, the office, every available inch of work space is pressed into service during the busy season.

A. B. Dick offset press, folder, and cutter used to print, trim, and finish the books we sell.

Envelopes in their thousands and sales literature in their tens of thousands help to make our operation BIG and SUCCESSFUL.

155.000 sheets of paper stacked and ready to print the tenth (revised and illustrated) edition of this book. Two to eight tons of paper, including envelopes, are consumed each year in the operation of our publishing business. Another ton or so in the operation of our worm business.

Paper is a vital part of any business, and as business grows, so will the paper-work and the weight of paper consumed.

The author's desk–organized chaos...most of the time. At other times it is just as disorganized as it looks.

Names, names, names--all neatly filed by state, alphabetically, keep our sales performance at our fingertips, and in touch with our customers at all times. Correspondence is coded on these 3X5 file cards because actual letter filing would consume much space not available in our home-office.

Is there money in earthworms?

They paid for the $5.000.00 (five thousand dollars) composer pictured above, on which this book was composed. The books, incidentally, just about break even, possibly even lose some money. But it is difficult to sell worms unless the buyer knows how to take care of the worms after purchase.

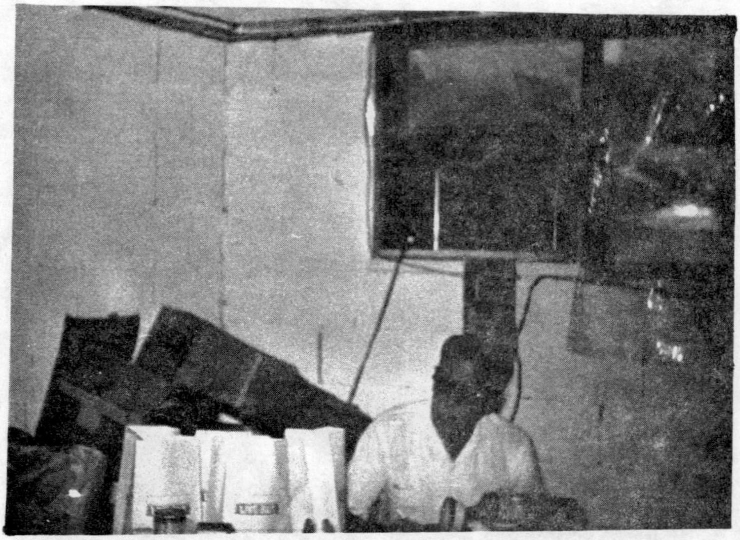

One of the farm offices where shipping records are kept by date ordered, what, how many, and when shipped. A duplicate record is kept in the main office.

Crated and ready for delivery to the postoffice or airline. This flexible wooden crate is the most adaptable package ever discovered for hauling live bait. The slats and open spaces at top, bottom, sides, and ends make it almost impossible to stack them in a way to exclude air.

Customer's cars line up to haul away worms on "pick-up" day. Thursdays are reserved for drive-in and pick up. Sometimes a million worms are reserved for this one day when people come from up to a thousand miles for a day or two of fishing and business.

great deal of variation in both method and route of movement. Mail loads may be shuffled to take advantage of lightly loaded baggage runs. By the time this book is printed, rail movement of mail may have ceased except in isolated cases, then routes may be shuffled about in order to keep minimum weight-mile contracts in force. Rural deliveries are especially troublesome because of lack of proper facilities, know-how, and interest by rural carriers who cannot or will not protect large packages from the elements. Cartons of worms are often placed in the rear of station wagons where they are exposed to direct rays of the sun shining through glass windows--and this may be deadly for any animal, no matter how well ventilated or packaged.

The rural carrier has dozens of stops with very short runs between. This may not allow for sufficient movement of air over the packages. The packages may overheat from radiation absorbed through metal car tops. Some carriers have car top racks in which large packages are placed. In wet weather these may be covered with a canvas or with a metal snap-on cover. Obviously, your worms are going to suffer if these conditions prevail, no matter how well you package them, label them, or otherwise care for them while they are in your control. Some conditions are simply not reversible, and there is little or nothing that you can do about it except do your part perfectly and hope for the best.

Temperature and humidity vary a great deal from one section of the country to the other, and even from day to day in the same location. This will seriously affect cooling of your worms through evaporation and movement of air through the package. Evaporation is the only method of reducing the temperature of the worm below that of the surrounding air unless ice cooling or mechanical refrigeration is available. Evaporation occurs only where there is sufficient movement of air through the package to remove moisture. And evaporation can only occur when the air contains less moisture than the packaging material.

Every shipper of worms should know something of both humidity and temperature in the area into which the worms are being sent as well as in the area where they are being packaged. In the case of long-distance shipping, some knowledge of routes (which may be variable, as noted above) and in-between temperature and humidity may be essential. One shipper sends worms from the east to the west coast (or vice versa) with few problems. Many will not attempt deliveries outside the fourth or fifth postal zone from their farms. The first has few losses, never insures. The second loses up to 15% of the worms he ships unless he uses air transport. They both use exactly the same packaging materials and packages. Obviously one knows something that the others do not, or practices a different approach to the problem.

A study of the various operations revealed that No. 1 trained packagers and either closely supervised the operation or did it himself. The others made it part of the duty of each picker to package their own worms as they brought them in from the beds, without supervision except for occasional checks...and in one case the grower was absent most of the time. That might be the difference between a profitable operation and an unprofitable or less profitable one. A good worm picker is not necessarily better qualified to package worms than a good jet pilot is qualified to repair the plane he flies. Only a period of careful study and training makes a good packager.

The chart I will give is effective for the red worm varieties as well as the African nightcrawler. (I group all the fancy-named bait worms together under "redworms"--brown nose, hybrid, English, Egyptian, red gold, Georgia brown, striper, curler, tiger, etc. as they are but four or so species so closely allied that only an expert can tell the difference.) By using an extra 10% of water in every case, the smaller kinds will actually improve their size in the package if not done in by poor handling.

The African worm (African nightcrawler) can be shipped almost completely dry-that is, in peat with just a trace of moisture in it, but they will arrive in a shrunken lifeless condition. The difference is that they will recover, if placed in wet peat or other good bedding, and within 24 to 48 hours will be as lively and large as ever. Many shippers deliver these worms in such a condition that an inexperienced person would think them dead, and without proper immediate care they will be, but all they need is a moist habitat. The trouble is that too many shippers do not bother to instruct their customers in advance, with resulting loss and disappointment. With proper understanding between seller and buyer, a great deal of money may be saved in the transportation of African worms--but here it is exceptionally important that the small amount of water used be allowed to soak into the peat for at least 24 hours before use...48 to 72 hours is better. Use only 25% of the amounts given below, but be sure and still take into consideration the southwest desert areas, where evaporation is faster and dehydration more likely to do permanent damage--for these regions add 10% (use 35% of the charted amounts).

From Florida to east of Mississippi but south of Mason-Dixon line: 10 pounds of peat to 4½lbs. of water. From Florida to west of Mississippi but EAST of Arizona and south of a line through Kansas City, Kansas: 10 pounds of peat to 5½ lbs. of water.

From Florida to entire west coast, Arizona, New Mexico, Utah, and Nevada: 10 pounds of peat to 6 lbs. of water. From Florida to north of Mason-Dixon line (Mason-Dixon line is roughly a line from Chesepeake Bay to nearest point on Ohio River following that river to the Mississippi), New England, midwest and northwest: 10 pounds of peat to 5½ lbs. of water.

Georgia and Alabama: Same as Florida.

The central location of Missouri, Arkansas and Tennessee permits narrowing of mixtures to two examples: all north and within 300 miles east or west: peat 10 pounds, water 5 lbs; 300 miles and over, south, east and west: 10 pounds peat, water 6 lbs. Other mid-west and central south states: same as above.

Texas and Oklahoma are usually about the same summer temperature and humidity and you must add more water because of the prevailing low comparative humidity: within 400 miles, peat 10 pounds, water 5½ lbs. Over 400 miles except west: peat 10 pounds, water 6 lbs. West, all shipments: peat 10 pounds, water 6½ pounds. NOTE: In these areas peat allowed to stand in open containers for more than 12 hours after presoaking is almost certain to need some additional moisture before the worms are packed-this is especially true if there is any movement of air.

Southwest, New Mexico, Arizona, California, Nevada and Utah: To all areas outside local deliveries: 10 pounds of peat to 6½ lbs. of water. California to west of rockies and north to Washington: Peat

10 pounds, water 5½ lbs. To southwest and east of rockies: 10 pounds of peat, 6½ lbs. of water.

Northwest (Oregon, Washington, Idaho) to all areas except southwest: Peat 10 pounds, water 5 lbs. To southwest: Peat 10 pounds, water 6 lbs.

Louisiana and Mississippi west to New Mexico: Peat 10 pounds, water 5½ lbs. To California and west: Peat 10 pounds, water 6 lbs. To east and northeastern states: Peat 10 pounds, water 5 lbs.

East coast states, Maryland and south: Same as Florida.

States north of Mason-Dixon, but east of Mississippi, all New England, New York, Pennsylvania, etc.: Peat 10 pounds, water 5 lbs. except to southwest: Peat 10 pounds, water 6 lbs.

States north of a line through Kansas City, Mo., but west of Mississippi River; all northwest, Dakotas, Montana, Wyoming: Peat 10 pounds, water 5½ lbs.

The chart above is effective when normal summer temperatures and humidity prevail in your area. On the average, the chart will be most effective during the hot summer months, but even then daily variations may be expected in both temperature and humidity. Variations will usually balance themselves so that no change need be made. During early spring and late fall, when cool nights are the rule, but daytime temperatures may be as hot as normal summer days, the cooling effect of night temperatures may be taken into consideration to save some money on transport costs by reducing the amount of water. But keep in mind that the temperature change in your own locality may not be in effect in the area to which you are shipping. This is especially true of from north to south and from the cool northeast or northwest into the southwest.

All local television and radio stations, and daily newspapers, feature a national weather report giving expected minimums and maximums of temperature and normal humidity. Learn where to find these services and use them-a few minutes a day devoted to weather reports might earn you several dollars a day in postage and loss prevention.

For winter delivery of red worms and other types that may be delivered in cold weather, use 20% less than the minimum amount of water called for in the chart. Very few people attempt to deliver African nightcrawlers during the winter, or through cold areas, but some do, and the very minimum of water is effective below about 65 degrees. In fact, the less water the better, because water is a good conductor of cold, and dry peat moss may help to insulate the worms from cold that might otherwise kill them. This is of course true for all worms shipped during the winter, but the red worm types shrink drastically, and may die without a moist environment.

Please, always keep in mind that the figures given here are not arbitrary: Your own requirements may be better served by a variation. But search for the perfect combination of moisture compatible with delivery of a product in best condition at least expense. The saving of weight is important, but it is one of the costs of doing business in the worm industry; conservation of customers is the first requirement. Experiment with small shipments and make notes on the results--if you can depend on the receiver to make an honest report.

Peat that has been stored for a long time may have absorbed some moisture in humid climates. This can be felt with the hands and allowance made for it, you will be surprised at the short time that will be required for you to learn the "feel" of properly fluffed and prepared peat. Every good packager develops this touch quickly.

Two people, buyer and seller, who understand exactly what is occuring at the shipping end and what should be done at the receiving end, can save a good bit of money by shipping all kinds of commercial worms with about half the moisture called for here. If the receiver has bins in a cool (not cold) place, with wet peat presoaked and ready, the worms may arrive small and stringy from dehydration, yet be fully recovered and ready for sale in 24 to 48 hours. The peat or bedding in the bins should be wet enough that a drop of water can be squeezed out of a handful of it. The worms should be fed lightly or not at all the first 12 to 24 hours, or until they have crawled into the bedding. But they should be fed as soon as they have crawled into the bedding, otherwise they may begin looking for food and escape from the bin.

The dry packaging methods described are especially effective with African nightcrawlers, and they can be hazarded in colder weather because of the insulating quality of the dry peat. Africans should not be taken into an air-conditioned or cold room for unpacking, *but should be acclimated by permitting them to remain in the package until they have gradually lost heat that may have been absorbed in a warm postoffice, truck, or automobile.* A sudden change of temperature may kill African worms or send them into such a state of shock that they cannot crawl into the bedding, therefore may dehydrate and die. And it may be necessary to feed Africans immediately with a light sprinkling of lay mash or fine ground corn meal as they will sometimes crawl out of the bin while searching for food.

Of course when shipping to a new customer or one who may not have the facilities or know-how for conditioning worms, it is better to avoid trouble by including a full quota of moisture. In this way you will avoid many complaints. Previous correspondence is the only way to be sure that you can cut corners without trouble.

By the same understanding, cups and prepackaged worms may be shipped almost completely dry or with only about half the usual or necessary amount of peat. The receiver has presoaked peat, opens each package, shakes out the old peat and replaces it with new, fresh material. In this way much weight may be saved and the worm reaches the consumer in far better condition than even when prepackaged perfectly at the source. Feeding at this time preserves the worms for a longer period, and often they will grow larger in the package.

PREPARING THE WORMS

To successfully package worms for shipment, they must be prepared for packaging. Preparation should begin in the beds before the worms are removed. This applies to some extent to all types and to a greater extent to some. Again we will recognize, for purposes of easy identification, two species of cultured worms: Red worms (called variously by the numerous names given previously), involving 3 to 5 species, and African nightcrawlers (Eudrilus Eugenie). The same instructions apply to all of the first category and generally, though different in minor aspects, to the second. Wild worms, field dug worms, leaf worms, drift worms, garden worms, pink worms, grunt worms, dew worms, nightcrawlers, nightwalkers, swamp wigglers (grey nightcrawlers), and others will be dealt with separately because of the specialized nature of their ecology.

It is essential to good packaging that when worms are placed in any type of package that they be free of any active organic matter that is not placed there to serve a specific planned purpose. This means that they should be free of any pit bedding or vestiges of feed that may cling to them or be transferred by the picker's fingers in the gathering process. Both the bedding and feed are composed of such a diversity of materials from one geographical area to another that this might become a separate problem for individual study by the grower concerned. Depending on the type of bedding, the type of worm, the composition of and method of feeding, preparation for picking should begin some days in advance of actually removing the worms from the bed.

Under normal conditions, worms can be removed faster and with a lower mortality rate to young worms if the bedding is slightly drier than is considered best for normal growth and reproduction. This is true of both red worms and African worms. We should plan to water lightly at least one day in advance, and in the case of red worms that may be grown in deep beds with a heavy active organic material such as cotton bolls or motes that have a high rate of absorption and slow rate of evaporation, or manure that has a still lower rate of evaporation, we might have to water lightly or not at all for several days in order to prepare the bedding for fast removal of the worms. Of course speed in removing the worms is not the sole reason for reducing moisture, but it is one of the main reasons. Obviously if picking speed by workers can be increased by 15% or 25% you are saving this percentage of the labor cost, and this is a very large item in any worm farm.

Reducing the amount of water in the bedding will cause a movement of the worm population. Sometimes this is good, sometimes it is bad, but on the average, during hot weather, when most worms are picked and the bedding is loose and fluffy from much digging and stirring, it is good. It will cause the worms to spread out in the bedding, rather than be collected in masses around areas of feed and moisture. This more or less even spread of worms makes it easier for the pickers to sort them into the proper sizes as they are uncovered, and may prevent the uncovering of many more worms than can be gathered before they crawl back into the bedding. Once worms have been exposed and have crawled back under cover, it is almost impossible for pickers to find them again until the bedding has been watered, the worms fed, and they have collected at their normal depth, which will vary with the season, food supply, and moisture. This is one of the reasons that pickers can seldom find more than 20% to 30% of the total worms available in one picking--but of course this has its good points, as the same bed thus continues to produce worms for periodical removal every few days rather than for only one or two crops in a season. This is especially true of the African nightcrawler.

Depending on the type and amount of feed used, this should also be reduced to some extent two days before any scheduled picking of that bed. If manure is used as feed, enough time should elapse to permit the worms to practically clean it up before they are picked. In the case of dry mashes and ground grains scattered on peat and the soilless types of beddings, the time element becomes less important and can be shortened to one day; but the worms should be fed lightly the day before picking begins so that they will be sure to remove all of it. This is not only to avoid picking up feed along with worms, but also to avoid mixing the feed into the bedding. Any feed mixed into the bedding is subject to subsequent souring and possibly heating, attracting flies, mites and other undesirable insect pests. This is especially true of the African nightcrawler as they are surface feeders, and any feed mixed into the bedding is not likely to be found and eaten by them, as it would be by other types of worms. African worms are also more agile at escaping than most others, therefore fewer of them are found and picked in proportion to bedding population than almost any other. Of course, being surface feeders, they are easier to gather together in suitable numbers for a second picking than most others.

African worms can be picked every other day in heavily populated beds. Of course each picking produces fewer and fewer worms, and eventually they must be given several days of "rest" to permit recovery of size and for growing out by younger worms. A good watering and light feeding immediately after picking, then a light watering and light feeding the following day prepares them for picking again on the third day in very heavily populated beds. Beds in which the worms are more thinly settled should not be picked more often than once every several days, otherwise it will be cleared of breeders, with gaps in future production of young worms. And a long period of time may elapse before the small worms are large enough to pick. The main hazard in this is that proper watering and feeding may be neglected in an unproductive bed, resulting in loss of young worms either through crawling or starvation, or both.

The red worm types should only be picked every several days--five days to two weeks. This gives time for better development of both older and younger worms and results in heavier, livlier worms, more acceptable to customers. If beds are picked too often, the population may be soon reduced to the small, young worms popularly called "bedruns," but which are really culls, or young unsuitable for quick rehabilitation by a dealer. The older the worm, the larger it is, on the average providing it has sufficient food and water and good bedding. After the clitellum or "collar" is formed, worms are very susceptible to the fattening process. But only age will produce the collar and potential tissue formation that will absorb "fat." The age must be gained between or before pickings.

The usual method of removing worms from the beds is the good old-fashioned one of uncovering them with a short handled light fork or "digger," then placing them in a pint cup, counting 1, 2, 3, etc. under the breath until the proper number has been counted. There is no other method of grading properly, although counting a thousand then weighing will give a legal continuous count--that may be acceptable in the "bedrun" trade, but not in "large handpicked bait or breeders." Various types of traps and sorters are described in my book "Profitable Earthworm Farming," but on the average, the quickest and cheapest method is handpicking and counting. And there is no other real way to grade worms for size.

After the worms are placed in the cups, they must be cared for. They should never be left in sunny spots, or exposed to falling water. Many mysterious shortages have occured that could be explained by the worms simply crawling out of the cups after they are picked and left behind and exposed to the hot sun. This is especially true of tree shaded beds as there are sunny spots in this type of cover. And any of these things harms the worms, no matter what course the worms may take: and it is our purpose and intent to get the worms to the customer in the best possible condition.

Immediately after picking, the worms should be removed to the packing area. In most farms this consists of a table or work bench under a shed. The cups containing the worms are there placed in convenient rows-many growers use even numbered rows of ten or twenty cups, or with each row containing the number of worms to be placed in one bulk container, in order to eliminate counting each cup as it is packed. The area should be well lighted because the worms will react best in bright (preferably fluorescent type) illumination.

In the presence of bright lights, the worms attempt to bury themselves as deeply as possible in order to avoid exposure. Light inflicts physical pain-and may be the only kind of pain that a worm's limited nervous system is susceptible to; though they will react to other things. The worm's sensitivity to light is remarkable in an animal completely without eyesight or visual nerve of any kind, but it is felt in the loss of oxygen caused by a closing of the surface capillaries that permit them to breathe by absorption of free oxygen directly into the red corpuscles. With the closing of the surface capillaries, the worm begins to axphyxiate, or suffer from lack of oxygen. Under these conditions, it attempts to remove itself from the source of the light, and this it does by burrowing downward. Thus we make use of the worm's primal instinct in order to prepare it for delivery to the fisherman.

When the worm attempts to burrow downward in the cup, the dirt and bedding that was accidentally picked up along with the worms is pushed upward, where it collects on the surface of the cluster of worms. The dirt and bedding are then easily removed by any one of several methods. Some people turn the cup upside down on a sheet of tin, a tray or shallow pan, then pick the bunch of worms off with the fingers, leaving the dirt and some small worms and capsules which are later dumped back into the beds. Other people use the method of simply placing a finger against the clump of worms then whirl the cup once rapidly with the other hand, at the same time tipping it almost upside down. The easiest and quickest method for the individual is best, and a little practice will soon develop skill with any of the several ways you may choose. I use the tip and whirl method. The most successful worm shippers clean worms the second time, placing them back in the cup and repeating the first operation, or by picking off the small remaining amount of dirt after the worms have collected in the bottom again, before they are packed.

REMEMBER THAT IT IS EXTREMELY IMPORTANT TO THE CONDITION OF YOUR PRODUCT THAT THEY BE CLEANED THOROUGHLY BEFORE PACKAGING FOR SHIPMENT! If you have not been cleaning your worms, yet have been having no losses, consider the fact that you may NOT be delivering your worms in the *BEST POSSIBLE CONDITION*! And that you could supply a more satisfactory product with proper cleaning and care before packing. All of which may lead to more customers more satisfactorily served, resulting in a more profitable operation in avoiding replacements, or even to the point of reducing advertising costs because your old customers will not try a different source and may stay with you year after year.

If the bedding and pit dirt are completely removed before the worms are packaged, obviously you are not concerned with what may have been their feed or of what the bedding may consist, from fresh cow manure to new peat moss. It can do no harm because it has been properly disposed of. All types of commercially grown and wild gathered or dug worms should be cleaned in the same manner except that sometimes in cool weather they will not willingly burrow to the bottom of the cup or container into which they are placed. In this case a strong light burning closely above the containers will cause them to go down--or a fan blowing over them--but be very careful that the worms do not overheat or dehydrate.

PACKAGING THE WORMS

At this time a course must be followed that will depend to some extent upon a previous one. Each type of package has its proper time as well as its special method, and the method of packaging worms is almost as important as the type of package. Most worms are delivered in bulk packages for repackaging or conditioning and packaging by the dealer, as opposed to those prepackaged by the grower. At least this is true at the present time. As more sophisticated packages and methods of using them are developed, the reverse might very well be the case in a few years.

When worms are placed into a package where space is severely limited, they begin a movement restricted by the confines of the package wall. This takes the shape of an aimless wandering at first, then develops into a migration to a point of common interest; the most comfortable part of the container. The most comfortable area will naturally depend to some extent on the package temperature, humidity and movement of air. If it is very hot, the worms will all attempt to come to the surface, which will become a ball of worms in constant movement on top of the packaging material. If it is cool, the worms will find a common point of interest on or near the bottom of the container. In very cool or cold conditions, the worms will probably form a mass at the center.

In any case, we can take advantage of this movement to help eliminate the possibility that they will be damaged by pit bedding or feed that may have been included in the package. And the worms may have fed heavily shortly before being picked, and have enough feed in the alimentary canal to cause some damage when they are congested into a very small area. The feed may be eliminated as the worm moves through the material, so by using our ingenuity we provide for the proper dispersal of it.

A routine should be adopted at the packaging bench. Everything should be in order; preferably in the order of use: The container or tub of peat to the right, the space for the worm package directly to the front, the cups of worms to the left. Or vice versa for left-handed people--I once had an excellent left-handed packer but everything was wrong for me when I tried to help out. In this way unnecessary steps are avoided as the whole operation can take place from one position. This position may or must be changed from time to time as the work goes on, but it permits the greatest amount of work with the least amount of body and foot movement. This is important in a large operation where long hours and fatigue may contribute to errors in counts and methods.

For bulk packaging, the individual container should first have about one pint of barely damp peat moss placed in the bottom. This may be measured in a cup or the hand, and some packagers use a small metal or plastic scoop. After the bottom is covered to a depth of about one or one and a quarter inches, dump in three cups of 100 thoroughly cleaned worms. Measure another pint or so of peat on top of these. Dump in another three hundred worms and cover with another pint of peat. The top layer of worms must consist of four hundred or four cups of 100 each, to total a thousand in the package. Then another pint or so of peat. This is for bulk packaging 1.000 worms to the container, is used for bedrun African nightcrawlers and for breeder or fishing size red worms. In cool weather it may be used for 1.000 large or bait size African nightcrawlers as well, and may be used for 1250 or 1.500 large red worms ("wigglers"). In regard to the numbers given here: It is usual in the worm business to either count ten extra worms into each cup of 100 or place eleven cups of one hundred worms each into the package. The 10% overcount is standard procedure as it is almost certain that some of the worms will be undersize or injured in picking.

As to method, it makes no difference what type of container is being used except possibly in the quantity of peat used. Naturally if a small package is being used, you must use less peat and include fewer worms--the number of worms is never arbitrary, but the space occupied by a given number of a given size is, and neither very small or very large packages are economical for shipment through the mails. The first because of disproportion of weight of package and packaging materials to worms, the second because worms are often damaged by their own weight when large numbers are congregated together. We assume that a one gallon or five quart container is being used.

When the requisite number of worms and proper amount of material have been added to the container, there should be a small space between the top of the packaging material and the top of the container, after the one is lidded, or the other folded and stapled. This is to permit a small amount of internal evaporation and movement of air without condensation forming on the walls of the package.

In packaging larger worms, where a smaller number is included in each individual package, the same method of distribution should be used but more peat per worm should be used in order to prevent shrinkage. It is just as important that the package be well filled as it is that it not be overfilled. Peat packed tightly prevents proper circulation of available air as well as restricting free movement of the worms. The peat should never be packed in, or the package bounced or bumped to settle or smooth the material.

As soon as the worms are placed within the confines of the walls, they start the movement previously referred to. In this movement they leave behind the vestiges of pit bedding and feed, and it is distributed throughout the package instead of being lumped into one ball that may sour and cause damage.

In packaging smaller numbers in retail size containers, the worms may be placed directly on the bottom, the peat on top; the worms will find their choice of locations anyway, and placing them in a certain level will be of no help to them or you. In the case of retail size bags the peat should not completely fill the package, here, either, as it may be desirable to add small amounts of wet peat or during storage or conditioning.

When using large bags, such as the S-754 R.C. Bait Bag, the top should be folded over twice and then securely stapled. Plier staplers are available at any stationery or office supply store and are much more convenient for this use than desk type staplers. Smaller bags of many different sizes and types are available and the method of packaging is the same for all except the number of worms. Naturally 50 or 100 worms may be placed in a small package without attempting to separate them into smaller groups-but they should be cleaned thoroughly.

Rigid containers should be packaged in about the same way, with particular attention to the placement of the covers to prevent crimping the lip of the package. The African nightcrawler will crawl through the smallest crease in the top edge or cover of any cup. This is not so true of red worms, they will usually remain in the damp peat even when the package is demolished.

Prepackaging worms for shipment and delivery in small packages presents a problem of nutrition and storage not encountered in bulk shipments. The bulk shipments are intended to be opened at once by the buyer and the worms placed in proper bedding for conditioning and future sale. Prepackaged worms are intended to be stored on the dealer's shelves for an indeterminate time, depending on good fishing weather, the quality of fishing, and the dealer's judgement in preparing for a good or poor weekend or other period. It is almost axiomatic in the worm business that the dealer expects worms to last up to 3 weeks and more in storage with no attention on his part. It is just as proverbial that the grower stretches the truth a good bit when he tells the dealer that they will last two weeks. Neither is right unless a great deal of cooperation is given by each.

Worms can be fed in packages and can actually be grown out to larger size there, using the packages as small bins. In fact some growers are discovering that this is the cheapest and most logical method of growing out large worms. But the jostling in shipment and consequent probability that any type of feed will be mixed into the packaging material, with fatal results to the worms, precludes the use of the best methods by the shipper. The dealer must, if best results are to be obtained, do the feeding at his end of the transaction. This may not be true when worms are delivered by truck or station wagon to the retailer, but is true of any type of commercial carrier and parcel post.

And this is where the greatest benefit could be obtained by both parties to the transaction, if the grower would use, as part of his selling technique, the inclusion of enough small packages to repackage the shipment at the buyer's convenience. As pointed out before, this would save the grower money by reducing the shipping cost and would guarantee the fisherman larger, livelier worms. This again involves and presupposes knowledge and participation of both buyer and seller, and if this participation is not coordinated by knowledge and preparation, the results will not be satisfactory.

The retailer, by using information described in the chapter on package feeding in this book, can increase the size of his worms from 25% to 50% with a corresponding increase in their market value. The grower, by pointing out the advantages of this information to his customers, can increase his volume of sales in direct ratio to the extra volume *HIS* customer, the dealer or retailer, will gain by offering a better product.

When *ANY* improvement is accomplished in the worm industry, it reflects all the way back to the man who produces the worms-or all the way forward to the man who sells them to the fisherman.

In the use of pint cups and small rigid containers, the worms can be picked directly into them. They are then carefully cleaned and packaging material added on top of them. Loose, fluffed peat should always be used here, and if the worms are to be cooled out before shipment, should contain about 10% more moisture than called for in the packaging chart. In a very large operation the packager may place several dozen cups side by side in a large square, dump the soft fluffed peat on top, then sweep it around with a large soft brush. After the cups are filled and capped, the excess is swept off the table into a tub or container for use on the next batch. Any short cup that will save time, saves money.

Placement of covers on the rigid cups should only be done by an experienced worker because the slightest off-center pressure will result in a crimp in the lip of the cup. The vibration of transport makes the worms restless and they will search out the tiniest opening for escape. For the same reason, care should be used in folding and stapling bags, or in carefully adjusting the wire tie--although these are not nearly so critical as the possibility of crimps in rigid containers.

PACKAGE FEEDING AND CONDITIONING

The word "conditioning" is new to most people in the worm business, but for several years an isolated grower here and there has toyed with methods of furnishing a longer-lasting, better bait worm to his customers, and they rightly supposed that the final conditioning should be accomplished after the worms were taken from the beds, somewhat as cattle are fattened out in pens before being sent to market. After the African nightcrawler grew to popularity, a small group of devoted farmers in central Florida developed some specialized methods of "growing out" or "conditioning" the worms before they shipped them. They found that by proper conditioning they could command a premium price for the worms, have fewer losses in delivery and save a good bit of time and money in getting the worms to salable size, especially as earlier methods had involved moving the worms one or more times from one bed to another in order to get them quickly to suitable, and most valuable, size.

There are three situations in which the conditioning procedure can be accomplished. The first is that to be performed by the grower before the worms are shipped. The second is to be carried out by the buyer (dealer) while the worms are in storage waiting to be sold. The third is that accomplished by the dealer who may purchase cheap unconditioned worms for resale in his own retail outlet or who may purchase in large bulk quantities for conditioning and sale to other dealers, somewhat after the manner of range cattle buyers who purchase for their own feed lot. Normally the grower would be better equipped for conditioning worms because he should logically have the know-how. The middle man-dealer should have planned his operation in advance and should have both the knowledge and equipment. Many retail dealers who are on their toes competition-wise have installed conditioning facilities, or have cellar or basement cool room area where equipment may consist of only a few boxes or tubs, or even only shelves on which to store small packages, but which are effective for the purpose.

Naturally a better worm will reach the fisherman if they are continually conditioned from the time that they are taken from the beds until they are placed on the hook-and this is possible except for the time that they are in transit from grower to dealer. But when the dealer orders often and does not allow packaged worms to remain too long on his shelves, conditioning by the grower before shipment may be all that is necessary. A surprising development of the individual experimentation by the central Florida growers with the African nightcrawler, was that the ordinary red worm, or hybrid, was even more susceptible to improvement in small packages than were their larger cousins. This led to the paradox of some very large African growers buying red worms, conditioning and reselling them to some customers who wanted to obtain both types from one source. Due to the fact that the African nightcrawler grows much slower in winter, and the red worm much slower in summer, they periodically crowd each other from the beds to such an extent that they can hardly be profitably raised in the same beds. But there did develop some all-purpose growers who produce one kind in summer and the other in winter--though the extent of this opportunity has yet to be fully realized by more than a few people. A two year experiment proves that the future will see growers who will have both kinds of worms in their beds with specialized pickers who can sort them quickly and easily into the separate species, and count each kind as it is picked. Two pickers were easily trained to do this in the experiment, though other workers found it impossible.

The method of grading worms by hand without counting, then weighing, is gradually becoming more and more popular as a time saving device. This procedure would easily permit two types of worms to be grown in one bed, and be picked and graded at the same time by one picker, because it eliminates the problem of maintaining a double count. Extensive research has revealed that even small retail packages of worms can be weighed by an experienced operator more than twice as fast and much more accurately than they can be counted. Or the picker can maintain a count on one type and the other be weighed, with periodic count-checks to be certain that no serious errors occur. In research it was found that errors always favored the customer. In several hundred recorded checks, only half a dozen favored the grower. There is plenty of room for such advanced methods in the earthworm industry, and as costs increase, some of them will have to be adopted in order to maintain and increase the present high volume of sales.

The ideal conditioning room for any kind of worm is one especially built, insulated and cooled with refrigerated equipment and arranged for storage of worms in packages. The refrigerated air cooling equipment might not be necessary in the southwest or desert country because the evaporative rate is so high that the regular evaporative air conditioner might do a satisfactory job-and many of the high desert plains have cool nights during which rooms can be cooled by forcing air in at night with fans then closing up tight during the day in order to retain the cooled air and bedding; or by aerating water which is cooled and stored in underground tanks then recirculated through the building during the day. If such a building were properly insulated, a small air conditioner can work wonders for worms. In some northern and mountain areas the public water supply system furnishes water cold enough that it could be piped through a conditioning building and circulated through a radiator through which a fan is moving air. But modern one and two room air conditioning equipment is so reasonable in cost of purchase and operation that make-shift methods are hardly profitable, even for the most ingenious grower.

A man who is handy with tools or who has a room which can be remodeled into a cooling room, can prepare space for several hundred thousand or a million worms for $650.00 or less, including cooler, insulation, special wiring if necessary, and containers for growing out worms. The room should be equipped with shelves to accomodate the various types of packages that will most likely be involved. Some growers have prepared small boxes of sheet metal or marine plywood designed to hold five hundred or a thousand worms. Some growers also condition all worms in the shipping cartons for 24 hours before they are mailed. Most growers who ship in retail size packages condition their worms before they are mailed. These are usually packaged one day, conditioned over night, then shipped. One such package was examined, placed in the air conditioned storage room, allowed to remain about 20 hours, then was taken out and placed in a shed on an August day in Florida. The worms were allowed to remain in the shed for three days, then the package was opened and examined. The difference in temperature could still be felt with the hands in the damp peat. The worms were in much better condition than when packaged...which means that they would have arrived in the hands of a customer six hundred miles away still cooler than the surrounding atmosphere, and in much better condition than when they were picked out of the beds in Florida.

Worms, in a dropping temperature, have a tendency to remain inactive. Their pores close, thus withholding moisture from evaporation, and there may be some osmotic absorption of water from the damp peat surrounding them. Their length is generally reduced by about 10%, with a corresponding expansion of the diameter-and everyone is convinced that a short fat worm is preferable to a long skinny worm-at least they sell better. The effects of cold contribute to less movement, therefore less loss of intestinal filler (food).

As the peat moss warms, the worms begin to move about, and that is the ideal time for the dealer to be prepared to take advantage of their newly awakened apetite for food-this would probably be about the time he would normally receive them in the mail or by bus, and is the most opportune possible moment to add feed to the package, if they are prepackaged for sale, or to place them for conditioning in new peat in a cool place where they can be fed. The resurgent appetite will, at this stage, sometimes actually cause a worm to double in size and weight in three to seven days. And any worm grower or dealer knows that this is a condition highly desirable and valuable to him in additional sales.

Some conditioning methods are simple procedures to prepare worms for the heat and probable lack of sufficient oxygen that will be encountered along the route of delivery, for a cool worm needs less oxygen and emits less carbon dioxide. Other methods are fattening regimens in which feed as well as low temperature conditioning is involved.

The first is the simplest of all and is that being adopted by large numbers of growers who realize the economic advantages: The worms are picked and packaged for shipment, with the shipping carton or crate labeled and sealed. It is then placed in a cool room in which the temperature is maintained at 68 to 74 degrees. This permits cooling of the carton, the package, the packaging material, and the worms. As we have seen, worms thus cooled will remain cooler than the surrounding atmosphere for several days, depending, of course, on the temperature of that atmosphere, the temperature to which the worms were reduced, the rate of absorption from the surrounding air, amount of moisture contained in the packaging material, and insulation furnished by the carton and package.

Most growers consider the comparatively low temperature of 68 degrees to be a little drastic for the African nightcrawler, or for any type of conditioning storage for domesticated worms, but results prove that it is not too low for Africans, and that the red worms can be conditioned at even a ten or fifteen degree lower temperature--down to 45 degrees, in fact, and gain much longer package life by holding body moisture. All of which means a larger worm on the fishhook, more money in the till and more profits all around.

Many growers ship top quality bait, and it is so described, but for lack of proper care, it reaches the customer in such poor condition as to be useless, and a customer is lost.

A sudden reduction of temperature will result in damage to the

African nightcrawler. It has a fairly well organized nervous system and it can be shocked by a sudden temperature change to such an extent that one or more of its heart systems will stop, and subsequent relaxation of the artery wall permits formation of a blood clot that kills the worm in a short time. All growers are familiar with the "knot" or "blister" that forms on the "head" of worms that die from cold (or from many other things). This is nothing more or less than the swelling caused by the stoppage of one heart system (all worms have from three to seven "hearts") and the continued pumping of others resulting in expansion of the heart artery. The worm is much more likely to succumb to this type of shock on a reduction of temperature than on a rise, and the slow overnight absorption by package and packaging material prevents a sudden change.

On the other hand, sudden reduction of temperature seems to give the red worms and "wild" worms a shot of energy. When they are removed from high temperature and subjected to a sudden reduction to as low as 50 degrees, their appetite improves, their sexual activity increases, and a general quickening of all their reactions is noted.

It requires 20 to 24 hours for cooled air to penetrate bulk packaged worms in S-754 R.C. Bait Bags, 4 to 8 hours longer in rigid carton. The smaller the package, the quicker the penetration--and of course the loss of heat occurs faster when the package is removed from storage. Ordinary pint bait cups with six pounds of water to ten pounds of peat in the mixture, require a full 24 hours for thorough penetration. That is through shipping carton wall, cup wall, and peat. Worms with more or less water in the packaging mixture will require more or less time, with the drier peat requiring the longer time-and the dry peat losing its cooling effect more slowly.

Some growers pick worms and store them during slack periods, to be shipped perhaps several days later. If the time is no more than 72 hours, the regular packaging material and method can be used, but here an additional pound of water should be added for each ten pounds of peat because refrigerated air is deprived of some of its moisture in the cooling process and will remove some of that contained in the peat. Some growers use a rebagging for worms that have remained too long in storage: They use gallon ice containers for the storage, then prepare the worms for shipping by dumping them into a No. S-754 double wall R.C. Bait Bag shipper, then one pint of very damp peat is added on to replace moisture lost in storage...or they may be shipped in the gallon can by cleaning off some of the dry peat and replacing with peat presoaked and precooled. The peat dumped off can be resoaked for reuse.

A quick check system may be established by using ordinary household scales to determine exactly how much moisture has been lost. A few packages of worms can be weighed when they are stored then reweighed when it is time to ship. The difference in weight is the lost moisture. In any case, some damp fresh peat should be added because the worms will have worked through the old and it will be less bulky.

The same storing procedure may be used with prepackaged worms in small containers, bag or cup. Only in this case they are cleaned of the old peat in which they were conditioned, just as they were originally cleaned of pit bedding for packaging, and new peat added. The tub or container of prepared peat should have been precooled in the same room for at least 24 hours, and the worms repackaged in the cool room itself, thus preserving the advantage of the cooling procedure. When cooling activity is carried out, it is obvious that the very last minute permissable by the postmaster or other transportation agent is the time to take them for labeling or stamping. The worms are in transit before they begin to warm up, and may remain cool until arrival at their destination.

The second conditioning system is that in which a feeding regimen is carried out in conjunction with refrigeration or other cooling. In this case the grower usually makes use of a hard-wall open top container, which may be anything from a waxed foil-lined cardboard carton to a box made of marine plywood.

This type of conditioning presupposes a bin, bedding, and a feeding program to be carried out rigorously during the fattening and growing period. Water in these small containers may be even more of a problem, in some cases, than the food, because of the propensity of conditioned air to evaporate surface moisture faster than is normally the case. For this reason, a "slop" type feed is nearly always used. The term "slop" in this case refers to a mixture of water and any kind of feed. The food is put on the surface of the bedding or peat as a "pat" of the wet mixture, and any future wetting is done carefully with forethought to the fact that the bottom of the container probably has very wet bedding or peat, and only be dry on the surface. It is easily possible that a too wet mixture prevail at the bottom while the surface is too dry for the worm to properly eat the feed placed there for them. Some growers drill a series of small holes around the bottom sides of the container for drainage purposes as well as to permit a flow of air upward through the bedding.

The bin should be of a size that will hold a predetermined number of worms: 500 or 1.000 is the most convenient. Under actual working conditions this would translate into 600 or 1.200 worms in the container because of the possibility that some will crawl out, die, or be too young to condition properly. It is axiomatic that if the worm is not collared, if the clitellum is not fully formed, it will not grow out suitably to be called a large breeder or bait worm. Collared worms, on the other hand, have a remarkable facility for increasing their size rapidly--just as a full grown steer can be fattened very quickly, but as a calf takes its own time to become an adult. The 600 or 1.200 worms, after growing out, can be cleaned of the small bin material and packaged in peat and are ready to ship. They will need no further counting or grading, and it is likely that the buyer will receive a generous overcount which most growers feel he is entitled to and which is standard in the worm business.

Any number of things may be used as a bedding in the conditioning bin if refrigerated cooling is utilized. Even things that are not considered very good bedding for outside operations or large bin growing, are quite efficient in the conditioning bins. In Florida and some of the northeastern states, as well as Michigan, there is a local peat that is reasonably easy to obtain and cheap and which is used as a bedding or additive for growing all kinds of worms in large outside and inside beds. This is wonderful as a bedding in the cool room boxes. Various composts, such as those made from rotting leaves and grass clippings work well, as does Buss-Bed-ding. German or Canadian peat mosses are good either alone or mixed with dried ground cow, horse or rabbit manure. Experimentally, the mixture of 75% German or Canadian (or any local peat that may be available) and 25% of manure, were found to be most effective at the low temperature ranges. But less so at higher ranges because manure has a tendency to heat by production of native bacteria and enzymes. Higher temperature ranges (75 and up) require use of less active organic matter such as partly rotted leaves or grass clippings, Buss Bed-ding, sewer sludge, etc. mixed with peat.

Many growers use 50% clean fresh imported or local peat and 50% of their regular pit bedding in which worms are being grown. The pit bedding is allowed to dry thoroughly, and is then mixed with the dry peat and the whole soaked for 24 hours. It should not be very wet, not wet enough to squeeze a drop of water out of a handful after it has been allowed to soak overnight or longer.

The container is filled about four inches deep for 600 worms, the worms placed on top and permitted to bury themselves, then they are fed lightly to prevent subsequent crawling in search of food. In lower temperatures it is possible to crowd the worms more, with more and more worms occupying smaller space in lower temperatures.

Feeding is a regular chore here, with small amounts of food before the worms at all times, but with no large amounts that might become mixed with the bedding and be a danger in the matter of emission of acids or carbon dioxide that may be harmful to the worms. Daily feeding is considered the minimum, and some growers feed twice a day. The food is placed in a small wet pat in the center of the container on the surface of the bedding and is not spread or patted down; not thrown or dropped from a height, but placed so that any natural porosity it might have is maintained. Amounts of food are not arbitrary but are found by trial and error with different worms and different temperatures and different humidities and elevations. Keep in mind that worms eat less as they become larger and fatter, contrary to food animals, and eventually to almost stop feeding activity altogether as their body weight increases and discourages movement...though of course there will always be a few hungry ones.

Feed mixes are almost as numerous as the growers using them. The most popular at the moment is fine ground peanut hulls (65%), fine corn meal (10%), crushed limestone (10%), and alfalfa leaf meal (15%). All percentages by weight. A good mix is half sewer sludge, 25% chicken lay mash, and 25% alfalfa leaf meal. Sewer sludge should be fine ground for this use, and all mixes should be made wet because if mixed dry the heavier ingredients settle to the bottom. Most of these mixes will harden by dehydration after the pat is put on the bedding and must be crumpled with the fingers.

We must digress here and discuss some of the ingredients named. Sewer sludge may or may not be available in your area, and you may have trouble finding someone who even knows what it is. It is a common waste product of city sewage disposal plants--if they do not pump out the raw or wet treated sewage into a stream. The best man to ask about this is your local health officer--the county agricultural agent will know about him. Crushed limestone, calcium carbonate, seems to confuse more people than almost any other single thing I have mentioned in my many books. Well, it is used by every high school, baseball club, football team, etc. in the world. It is the stuff that is used to line the field. Its purest form is calcium carbonate and this is not irritating to eyes and cuts. And it helps to absorb acid or prevent it from forming in worm beds and feed.

After the food, of whatever kind, is wetted, it should stick together in a ball when picked up. This may require very little water but it should be made wet enough that it will not harden and be useless to the worms. Each container should have an amount about the size of a walnut if 600 Africans or 1.200 red worms, more if there are more worms. And it is important that each feeding should be cleaned up by the worms before the next is added--but it is just as important that food be present at all times. So they must be fed and cared for diligently after they are accustomed to the new climate.

There should always be a small area around the perimeter of the container that is free of feed to which the worms can escape in case of heating and the normal congestion that will occur. Some growers make the food mix very wet and use a dipper or large spoon to dispense it. A bit of spreading will not seriously affect anything so long as small portions are used, and the sludge, peanut hull or Buss Bed-ding mix will not ferment at all, or very little, though some mold may form. Molds are considered harmless, and usually, though not always, are.

Other grain and milled leaf mixtures are popular, depending to some extent on their availability in your area. Wheat bran, corn bran (shorts), peanut meal, various nut meals that are available in some parts of the country, chopped alfalfa and sludge, finely crumpled rabbit droppings and sludge; or old used bedding, rabbit manure and sludge in equal parts. Old used bedding and sludge become the bland part of the mixes in which they are used. The simple sprinkling of dry pablum or other finely crushed household cereals on the surface of the damp peat is an excellent and easy way of feeding.

If worms are to remain in the container for more than eight days or so, the old bedding should be cleaned off and new added... for some reason they will not continue to grow after about a week, but will take on new growth each time new bedding is added.

Method three we will call the impregnation system, and is utilized for packaged worms as well as for fattening them in small bins. Impregnation of the bedding or packaging material with liquid food is new enough that most growers have not really tried it, even though most of them should know about it.

The two methods of impregnation are: The impregnation of the packaging material with a liquid food and vitamin-mineral appetizer and the impregnation of an additive of either a sponge type dispenser or a covering material to be added or distributed over the bedding or packaging material.

Of the limited numbers of liquid foods available, peanut oil is probably the most universally available, satisfactory and, easiest to use. Corn and palm oil are usable in the same way if obtainable in pure form. Saff-flower oil has recently been tested and it appears to be the best of all tried to date. Peanut oil is universally available in supermarkets everywhere at a reasonable enough price to make it economically feasible to use in the amounts required for the purpose of feeding in the package. It is available and cheaper in 5 gallon cans at wholesale grocers.

Oil is used by impregnating dry peat, about 2 ounces to the pound of peat. Pour the oil over the fluffed peat and rub through the hands to distribute the oil over the surface of the fibers, then add the proper amount of water and presoak as in the chapter on Selection and Preparation of Packaging Materials. Both the oil and the water will be absorbed by the peat; the oil will trap some of the water, causing slower evaporation thereby keeping the peat in better condition longer. The worms will eat the peat, digest the oil, and retain a great deal of fat and moisture in the form of body liquids that are normally lost in congested packages.

The oil impregnation method is especially effective in conditioning and growing out worms if they are picked and stored in the shipping carton for later shipment. As they warm, the appetite is sharpened and they ingest the peat in large quantities in transit. Also the oil helps in maintaining body liquids and color. A better method for this purpose is to pack the worms in a container other than that in which they are to be shipped, but with impregnated peat, and condition for 48 to 72 hours. When they are repackaged for shipment the old peat is removed and newly impregnated, cooled peat is used and the worms have a continuous supply of food, which they do not have under present conditions and methods of packaging.

Buss Bed-ding, a commercial worm bedding manufactured and sold under patent by Buss Manufacturing Co., Lanark, Illinois, is excellent both as a bedding or as a feed for package feeding. It is not generally considered suitable for shipping or packaging material but can be used if extreme care is exercised in the wetting and pre-soaking procedure. With this material, the smallest amount of water compatible with a proper amount to sustain the worms is best. It is very easy to add too much moisture and when this occurs the material is such that heat accumulates rather than giving evaporative cooling, though the material itself does not generate any organic heat. It provides poor evaporation unless left dry enough for the fibers to separate and permit a flow of air. It is one of the few beddings to which food has been added in the manufacturing process, and in which the bedding itself becomes a food as it is broken down and digested by the worm. Buss Bed-ding works extremely well in the oil impregnation method if the worm is kept cool. The oil is soaked into the Bed-ding in the amount of 1 ounce to one pound of dry material, the Bed-ding is rubbed thoroughly to distribute the oil over the fibers and then the Bed-ding is barely dampened and allowed to soak overnight or longer.

Peat is prepared in the usual manner, the worms packed, then three to six ounces of Buss Bed-ding, prepared with oil and water, is spread over the surface of the packaging material (this is for bulk packaging of 500 to 1.000 African nightcrawlers or red worms). The package is sealed and cooled in the conditioning room for 24 hours, then shipped without further care. The worms begin to feed on the oil impregnated Bed-ding almost at once, if they are properly cool, but usually will not feed on such a mixture if the temperature is very high in the package (90 degrees and above). Although a hungry worm will feed to some extent in any temperature which permits movement. Worms of any kind feed very little or not at all when temperatures reach 90 degrees--this is one of the reasons that night feeding occurs primarily during the summer.

In small conditioning bins in cooled rooms, Buss Bed-ding may be used without impregnation but usually a small amount of lay mash, bran, or finely crushed breakfast cereal is sprinkled over it after it is dampened and soaked and the worms introduced. Sewer sludge may be added right to the Bed-ding, in any reasonable quantity, and the worms will eat it and the bedding. It should be mixed into the wet bedding to prevent filtering out as the bedding is mixed.

A small amount of vitamin-mineral mix will increase ingestion of food up to 100% in pre-shipment conditioning, but should not be used in packaging material in which worms are prepared for shipment. Ganatrex and Geritol are common liquid preparations for human use. The addition of one teaspoonful per thousand worms, diluted into the water with which the peat is prepared, will cause the worms to retain their size and color largely because they will eat everything in the package--and the package, if it is not a good one. Here is a place where temperature is the deciding factor. It is considered a waste to add any kind of food except vegetable oil to packaging material which has not been precooled before shipment, though small amounts of dry crushed cereal or baby food sprinkled on the packaging material does help.

For small packages of worms prepackaged for shipment and sale over the counter by the dealer, Buss Bed-ding is one of the finest additives. Ideally, it is added by the dealer at his end, but may be added by the grower before shipment and by the dealer after receipt, thus giving the worms a constant source of food--so long as it is not overdone, causing too much moisture to be added to the package. The Bed-ding should be dampened to the point that it will hold a form when it is squeezed or molded with the hand, wet enough to squeeze a drop of water out of a handful is about right.

For pint packages of 50 Africans or 100 red worms, use a pat of this dampened Bed-ding about the size of two silver dollars, stacked. This can be placed on the surface of the packaging material and it is harmless if accidentally mixed in while the worms are in shipment--but it may not be effective as a feed because African worms, particularly, and all worms under these conditions, will come to the surface and find the feed. Of course some sewer sludge may be mixed with Buss Bed-ding for this application, but no grain or other active organic feed. The bedding made from my Soilless Worm Bedding Formulas works equally as well as Buss Bed-ding, and is used in the same way.

Any worms shipped in any size package should be cared for immediately upon receipt by the purchaser. They are hot and agitated from much vibration and bouncing of transport equipment; there has almost surely been a lack of fresh air along the way, and very likely the moisture in the packaging material has evaporated to the point that the worm is now losing its body liquids to the dry bedding and dry air. This is the third province of care that can be extended by those interested in furnishing the fisherman with the best possible bait in the best possible condition.

THE DEALER SHOULD: Have a prepared place for storing and caring for live worms; Have some know-how that can be applied to the proper care of live worms. Too many dealers want worms to come into their shop "ready to use," and most of them do not realize that this may be a complete biological impossibility because a worm, in common with all of the live things on the earth, needs food, water and oxygen. And the grower, even those that do their very best, is limited to the very small amount of food and water that can be packaged with the worms. The nature of food is that it spoils and is likely to damage the worms instead of nourishing them; and of water that it evaporates and is lost to the air which must move over the worm in order to supply oxygen to body tissues.

Worms, fortunately, are able to exist and even maintain a large degree of vitality for several days without food--but not without air and water. The skin condition and color depends on water, as does the proper absorption of oxygen. And without skin tone a worm is dying. The grower cannot add too much moisture or the worm will die because of heat absorption, and because it is too expensive to ship. So it is up to the purchaser to add both food and water at his end--the sooner the better after the worms arrive.

A cold room with refrigeration down to 62 or 64 degrees is the ideal situation. If the store or shop is air conditioned for the comfort of proprietor and customers, it should be fine for all except "wild" worms. Next best is a cool, dark cellar or basement in which the worms can be stored while awaiting sale. Just ordinary storage, like fish hooks or lures, on shelves exposed to heat, and blocked from proper ventilation is not good enough for worms.

In one instance where a dealer complained that no matter where he bought his worms, or how much he paid for them, they would not live for more than two days, we foudn that he was giving them a prominent display--doing a wonderful job of selling bait worms, but stacking them in a large display window on which the full force of the afternoon sun shone for three hours every day. It is a compliment to the packager that they lived as long as they did under such conditions. This is just one example of the lack of knowledge, especially at the retail end of the business; though there is appalling lack of knowledge of packaging at both ends...A few minutes' conversation with the dealer convinced him that a stack of empty bait packages in the window for display, with the real bait carefully disposed in a drink cooler in the rear of the store, would do a better job all the way around. And subsequent events proved this true: his bait was in such perfect condition that he more than doubled sales in seven weeks from word-of-mouth advertising among fishermen.

The very best, most effective method for delivering the finest bait worms to fishermen is for a dealer to buy worms in bulk and have a few wash tubs or other small bins where the worms can be stored, fed, cared for, and counted into small packages daily as they are sold. Most dealers do something of this kind eventually but generally are woefully without the necessary knowledge as to proper bedding and feed, water and care. And their losses are too high because of this lack. Any commercially grown worm will store well in ordinary German or Canadian peat moss, and will recover size and vitality rapidly if the temperature is below about 74 degrees. Even old sawdust may be used if nothing else is available. Bedding made from my own Soilless Worm Bedding Formula, Buss Bed-ding, and other commercial beddings are good.

German or Canadian peat or Soilless Bedding and Buss Bed-ding, when used for storing worms, should be presoaked, but should be very wet, that is wet enough that a handful will produce a drop of water when squeezed. It should be well fluffed and worked with the hands and should not be packed into the tub or bin, but should be placed there loosely. About five inches of any of these in a No. 2 wash tub will accomodate 1.000 large African nightcrawlers, or about 2.500 red worms. Habitat and water are most important here, food is secondary because the worms gain some nourishment from the bedding. Buss Bed-ding and my Soilless Beddings need not be fed at all. Sawdust and peat contain no food values which the worms can exploit immediately, so need some feed.

Worms that have been "tubbed" can be fed many different things in many different ways. The best method is to follow previous instructions for wetting feed and using it in "pats." In this way, worms can be removed without mixing the food into the bedding. Bedding kept properly clean and free of food residue can be used over and over again, and Buss Bed-ding, or my Soilless Bedding, can be used continuously for a full season or longer, depending on how many worms are stored in it. It can be washed, dried and returned to use several times if feed is not mixed into it.

New peat should be kept mixed and wet at all times for counting out and packaging the worms as they are needed. The small sizes of R.C. Bait Bags are at their best here because the bag walls permit a constant flow of fresh air and give evaporative cooling. A small amount of sewer sludge or lay mash may be sprinkled on top of the packaging material if desired, or a small wet pat of any grain or cereal mixture placed there for the worms to feed on, but inasmuch as they probably will be going fishing in a day or two, food is not as important as proper cooling and a supply of fresh air.

Let me say here that African nightcrawlers should never be refrigerated and store well in any temperature in which humans are comfortable in shirt sleeves--72 to 82 degrees or higher, but not much lower.

Tubbed worms, or worms being stored in small bins of any kind, will sometimes attempt to crawl, especially if the temperature is above 82 degrees. This can be prevented with a small light burning directly over the bins, or in the case of a stack of small bins where lighting will be difficult, a small oscillating fan will usually prevent them from crawling. The fan, moving air over them, causes dehydration when they expose themselves, so they remain in the wet bedding. But the fan will also increase evaporation of moisture and, in the case of surface feeding worms, may prevent proper eating. In the lower temperatures, 72 degrees and lower, the worms seldom attempt to crawl.

A remarkable new "impregnation" feeding method used with considerable success, is that of soaking small pieces of cardboard, chipboard or ocean sponge (not foam rubber) in one of the various oils, then dropping one or more pieces on the surface of the bedding in the packages. The material should be soaked overnight in peanut, corn or saff-flower oil, then a piece the size of a silver dollar used for each cup or for 250 worms in bulk packages. In order to hurry the process, the cardboard bits may be "fried" at low temperature, then permitted to drain--but the oil should not be over about 180 degrees for the frying process.

The future of the worm business may depend, to a great extent, on how well we who are in it carry out the necessary research and testing of every possible avenue of improvement of the product. Added costs in every department, from feeding to transportation, are becoming a serious factor in production and distribution of bait worms, and methods of increasing size while in transit or at the dealer's shop will obviously eliminate the necessity of paying transportation costs on heavy worms. Changing transit methods might eliminate the long-distance shipper in the near future unless we work out methods of combatting slow movement and careless handling. This book is one step.

NATIVE NIGHTCRAWLERS, DUG WORMS, ETC.

The "wild" worms are those gathered from the endemic population and not ordinarily raised commercially. These are the Canadian dew worms (nightcrawlers or nightwalkers of the northern U.S.), leaf worms, drift worms of the northeast, pond and river worms of the midwest, dug worms of many areas, pink worms, grunt worms, and swamp wigglers (grey nightcrawlers) of the south. Practically any area except the sand desert of the southwest has some kind of native worm available when it rains or in the spring when the earth is damp. Most of these worms are found in soil or water in which the temperature is usually much lower in summertime than the air. The worms move up and down in the soil in a pattern directly related to the moisture and temperature of their environment.

In times of drought or excessive heat the wild worms may "estivate" or become dormant in a deep burrow where they remain until minimum conditions are restored at or near the surface. This would indicate that these worms have considerable recuperative powers, as well as resistance to a hostile environment. But their resistance is strictly limited by ecology: They have no resistance whatever to excessive temperature--and in some cases excessive temperature may be as low as 45 or 48 degrees F.

Most North American wild worms will live very well for some time at temperatures up to 60 or 65 degrees, but for long storage this must be reduced to as low as 38 degrees. The northern nightcrawler (dew worm) must be stored at no higher than 42 degrees over long periods and can only be shipped if the package is somehow cooled. This is done by various ingenious ways, usually by wrapping dry ice in a package of newspaper (to protect the worms both from the "burn" of the "dry" ice and the fumes--carbon dioxide--of its thawing), then sealing the package fairly tight to prevent loss of the cold air. They are usually shipped in small lots by air; large lots are hauled by refrigerated truck. Obviously a worm so cooled would not need food.

It must be assumed that no wild worm can be successfully stored or conditioned without artificial cooling. Nightgathering of worms of all kinds, and digging field and garden worms, is accomplished during spring days when temperature and moisture conditions are at maximum for the worm to live near the surface and come out, or partly reveal itself, at night to feed or mate. God or dew may contribute to high humidity in order that native worms may live partly exposed to the air at night. This natural environment must, to some degree, be duplicated in order to store or reproduce any wild worm.

The "pink" and "grunt" worm (or any other wild worm) of the south does better at slightly higher temperatures but will keep longer under the lower temperatures. A discussion of the proper packaging of any wild worm will devolve to a series of pointers on proper storage temperature because neither habitat or moisture are critical except that the worms must have enough packaging material to cover them and it should be damp enough to prevent loss of body liquids, or replace that lost by evacuation. In many cases the northern nightcrawlers are hauled in open trays without benefit of packaging material or bedding of any kind. They are cooled by icing or by night travel. Their body structure is such that they are not encouraged to crawl under any condition (up the sides of cartons or trays) as are the red worms and tropical nightcrawlers.

Presoaked peat moss, Buss Bed-ding, aged sawdust and the normal mixes used for packaging any kind of commercially sold worms are also fine for all types of wild worms. But in the midwest a sort of "gumbo" or muck is used for packaging some of the "river" worms and others. Excessively wet peat or other materials described here would work as well, inasmuch as temperature is the controlling factor in keeping them, anyway.

It is a sad and shameful commentary in this respect that worm dealers and wholesalers will spend hundreds, even thousands of dollars a year on advertising, driving about finding customers, trying to find a satisfactory package, and on almost any phase of dealing with live worms than they will with trying to find satisfactory means of keeping it or delivering it after they have the worms and the customers. Customers are lost, money returned, and even mail privileges are revoked because of failure to solve delivery problems.

Customers should be instructed, preferably with a printed sheet included with the worms or mailed in advance, that all worms are susceptible to heat and dehydration and that proper care should be exercised in the preservation of his bait until it is used or sold.

Many, in fact most, fishermen use some type of cooler for drinks or food on fishing trips, and in the case of wild worms of any kind, this cooler is ideal for keeping his bait provided that the package is not allowed to come into direct contact with the ice or become immersed in water resulting from the melting.

Proper instruction all down the line, from picker or grower to fisherman, is imperative to the successful use of wild worms as bait. Let us all work to make it as easy and pleasurable as possible for the guy who pays the bills--THE FISHERMAN.

THE LARVAL WORMS

The larva are not properly worms at all, but insects in one of their pre-pupal stages. These are widely used throughout the world as bait but very few kinds are sold on a commercial basis. Usually they reflect the work of local collectors who specialize in the natural forms available at different seasons from the water or foliage. Surplus supplies are sometimes sold to processors who preserve them or have facilities for cold storage.

Shipments of these forms are not any different than that of earthworms except that time is a most important factor, as well as temperature. All insect larva will, in time, metamorphose into the next larval stage, where it may not be suitable as bait, or into the adult insect. This makes it imperative that they reach the market in the shortest possible time. By the same token, all larval forms are extended by lowered temperature, with the possible exception of the helgrammites.

The commercial larva most widely advertised, and most commonly raised, are "meal worms," larva of the prionus beetle, wax worms, larva of the bee moth or wax moth, "mousee" grubs, maggots of one of the blow-fly family and "silver spikes," maggots of the common green blow-fly. Some European corn borers, "stem" worms that grow from eggs laid on the stems of aquatic plants, "catalpa" worms that may be anything from one of the army worms to a webworm or ordinary caterpillar, and helgrammites, larva of various may flies and dragonflies are offered for sale on the market. Many of these, notably the corn borer, have built-in tools for major carpentry and must be confined in metal.

Normally most of these are packed with nothing except moist material from their own environment. The mealworms are packed several hundred to a pint cup with some bran or dry corn meal with a small piece of apple or potato for moisture. The others must be dealt with according to their various requirements of environment. The most important thing in all these baits is to prevent metamorphoses by maintaining a low temperature. Meal worms, corn borers, helgrammites, etc., can be stored in an ordinary household refrigerator at about the same temperature as green vegetables.

It is imperative with these short-lived creatures that both the dealer and the fisherman be instructed as to proper care, and only a specialist who knows the proper methods should attempt to handle them. Even the use of these exotics is strange to many people who would be happy to give them a try if knowledge were volunteered in the proper procedures. The helgrammites, especially, are a favored and excellent bait in many northern streams and lakes and they are hardy and long-lived if properly cared for. And they are plentiful enough in early spring to be extremely profitable if gathered and sold on a route out of a properly constructed storage facility.

A little study will indicate the species and seasons for demand of insect larva in your area, and this could lead to an extremely profitable sideline that would pay far higher wages than any mill or factory. An old refrigerator might be the nucleus of a $10.000 a year business--I know of at least three of four of these...but please do not ask me for names and addresses, these people have worked out extraordinary deals for themselves and are very competition conscious, as well as individualistic, and those who have been willing to talk to me at all have specifically asked that names and addresses not be divulged. The leads are there, however, for any person with initiative and imagination. One thing, some of them advertise in the outdoor magazines--that is the way I found them.

A FINAL WORD

It seems that the end of a book always comes before it is finished. The writer finds many things left out that he intended to include--and even some things that should have been included. But unfortunately the space is limited. Any book must be planned in advance as to time, publishing costs, approximate contents both as to subject matter and number of words. Research items must be foreseen and arranged for, interviews arranged and disclosures tested (some growers will mislead either through ignorance of the subject or maliciously), and even page or chapter headings planned for space requirements. The effort expended in gathering the information contained here and in getting it in book form was almost impossibly complicated.

Every research item must be proven under actual operating conditions--and in many cases it will fail you for some unknown (to you or me) reason after innumerable tests. Statistics are worse than useless, they are misleading, so figures must be expanded to live words, and messages confined to facts and facts repeated at least once. The truth is not always easy to arrive at, even when it can be seen, because there may be a maze of unrelated truths to wade through. Simulations will not prove anything in a live bait operation because simulations are always exact, while conditions under which you will operate will be constantly changing.

One of the main facts that has changed since this book was started is the Postal Department's handling of the mail. At that time mail was mostly handled in baggage cars where mail sorters worked and where conditions were comfortable for them. Now hardly any mail is sent in baggage cars moved by locomotives attached to passenger trains. First class is sent by air while all other kinds go by truck. It may be the so-called "piggy-back" truck loaded onto a flat car and hauled in a train of cars, but it goes by truck, nevertheless.

One or two growers who examined the manuscript before final editing expressed disappointment that more marketing information had not been included. But here limitations are no less effective than in other departments, and that marketing, in the final analysis, is proper presentation of a product properly grown, conditioned and packaged.

Marketing earthworms for bait has never been considered a major problem once the age and size is acheived, and the growers who have been able to produce a worm that may be considered the ultimate in quality, turn away more customers than they serve simply because the production of a large number of bait worms and the production of desirable large bait worms are not compatible. I will say without equivocation that those growers who produce good quality bait worms produce many fewer than those who just produce worms. And that those who produce first class worms delivered in first class condition are never in need of a sales campaign. Probably the knowledge of a good source of bait travels faster by word of mouth than any other one bit of information possessed by man, because usually fishermen are gregarious and good talkers as well as good listeners.

"Secrets" revealed in this book will be resented by some experienced and established growers who consider the value of research or accidental findings to be the sole property of the finder. The writer probably possesses as many of these "secrets," discovered by his own efforts as well as by the efforts of others, as any man alive and yet I have never taken the position that such professional knowledge is private property, to be withheld or dispensed at will. It is my view that only politicians and criminals (this is by no means a categorization of the former with the latter) have secrets. The findings of any scientist should, upon proper demonstration of factual proof, become public property, to be freely used and profited by, if profit is possible without victimizing others, without let or hindrance. If the scientist can profit from the dissemination of that knowledge, in the form of a book or essay, so much the better, because any true craftsman of the science of discovery will use the proceeds of one to go on to others.

The composition of books and themes is almost the sole stock in trade of any researcher below the status of inventor. The true scientist can only market the product of his mind, and fortunately we have the copyright laws, with the stiff penalties for plagiarism, to help him protect his writings. The inventor has a mechanical product to offer as a result of his research, with proper patent laws to protect him. And all this is the reason that no true "secret" can exist in science. If it is a secret it is not knowledge and is not marketable.

There is still one main "secret" to be discovered and disseminated in the field of growing earthworms--and that is how to convince people that they must learn from others through reading. Probably not more than 50% of any book ever reaches the mind of the man who bought it, and very likely not more than 25% of worm growers ever buy a book worthy of the name. "Booklets" yes! The smaller the publication, in most cases, in the worm business, the higher the price...and the more flamboyant the advertising and promotion to sell it--because the tremendous profit enjoyed by those who sell such trifling sheets permits spending money on overselling.

But ANY source of information is better than none. Everybody needs a starting point--as it is so truly said "a journey of a thousand miles must start with one step." We can only hope it is a step in the right direction.

Even if you do not agree with the conclusions drawn in this book, PLEASE READ IT. The information is factual and valuable.

Good luck, good sales and good fishin.